Hopp

Plastics Recycling in Transformation

Matthias Hopp

Plastics Recycling in Transformation

Technological, Economic, and Political Aspects

HANSER

Print-ISBN: 979-8-950184-12-3
E-Book-ISBN: 9798950184321

Bibliographic information of the German National Library:
The German National Library lists this publication in the German National Bibliography; detailed bibliographic data are available on the Internet at http://dnb.d-nb.de.

Vilshofener Straße 10 | 81679 München | info@hanser.de
www.hanser-fachbuch.de
Lectorate: Dr. Mark Smith
Production management: Frauke Schafft
Cover concept: Marc Müller-Bremer, www.rebranding.de, München,
Cover design: Max Kostopoulos
Cover picture: © gettyimages.de/Govindanmarudhai
Typesetter: in-house

Table of Contents

Acknowledgments

First and foremost, I would like to thank Prof. Dr.-Ing. Elmar Moritzer and Prof. Dr.-Ing. Volker Schöppner, who made it possible for me to develop the content for the habilitation thesis in my own research and teaching area, which forms the basis for this book.

In the course of establishing the new course "Plastics Recycling", I would like to thank in particular Dr.-Ing. Lisa Tölle and Dr.-Ing. Maximilian Scholle for their valuable support. In addition, we were able to develop an initial analysis of material flows in a working group, for which I am grateful to all participants. A heartfelt thank you also goes to Dr.-Ing. Frederik Mühlhoff, Leon Schäffer, Dr.-Ing. Laura Austermeier and once again Dr.-Ing. Lisa Tölle, who provided support in the implementation of research topics in the laboratory. In this context, the students who made a significant contribution as part of their student projects should also be mentioned. I would like to thank Anne Jäckel and Florian Löhr for their help in creating figures and diagrams, and I apologize retrospectively for the many revision cycles that I caused.

The fact that I was able to find the time during my work as chief engineer to continue advancing the habilitation is thanks to those who supported me in the established processes for budget planning and control.

Last but not least, I would like to thank my wife, who repeatedly supported and motivated me to complete the habilitation. Without her, this book would not have come about.

Preface

I am delighted to be able to present an English-language edition of this book, which was previously published in German under the title *Kunststoffrecycling im Wandel*. The topics of plastics recycling, the circular economy, and sustainable resource management are becoming increasingly important worldwide. It is therefore all the more gratifying to be able to make the content available to a wider international audience.

Although the scientific principles and technical contexts covered in this book are universally applicable, some sections refer explicitly to the regulatory framework of the European Union and, in particular, Germany. Readers outside Europe will therefore come across explanations of German legislation, waste law, and recycling requirements. However, these explanations should not be understood merely as a description of national particularities. Rather, they offer valuable insights into the way in which European directives are implemented in a Member State of the European Union.

In an increasingly globalized economy, material flows, recycling markets, and value chains are closely interlinked. By understanding the legal and economic framework, stakeholders outside Germany and Europe can better grasp why certain material flows are organized in one way or another and why markets for secondary raw materials are developing accordingly. The example of Germany clearly illustrates how regulatory requirements influence recycling infrastructure, economic incentives, and the availability of recycled materials. As one of Europe's largest economies, Germany is of particular significance in this regard, with its impact often extending far beyond its national borders.

At this point, I would also like to address an important terminological distinction. In the English-speaking world, the terms "mechanical recycling" and "chemical recycling" are frequently used. In this book, however, I have deliberately chosen to use the terms "material recycling" and "feedstock recycling" instead.

The reason for this, in my view, lies in the greater scientific precision of these terms. The distinction between the two recycling routes is based on how the molecular structure of the plastics is handled. "Material recycling" encompasses all processes in which the polymer chains are essentially preserved and the plastic is transferred as a material into a new life cycle. These include traditional mechanical recycling processes as well as certain solvent-based processes in which plastics are purified or separated without breaking down the polymer structures.

In contrast, feedstock recycling encompasses those processes in which the polymer chains are specifically broken down into smaller molecular units such as oligomers or monomers. These chemical feedstocks can then be reused to manufacture plastics or other chemical products.

I hope that this English edition will help to deepen international understanding of both the technological and regulatory aspects of plastics recycling, and promote professional dialogue between academia, industry, policymakers, and all other stakeholders involved in shaping a sustainable circular economy.

Dr.-Ing. Matthias Hopp

Summer 2026

1 Plastics and Their Recyclability

In addition to metals, ceramics, and natural materials such as wood, plastics are available to humans as a material. The distinctive property profile of plastics has led to a rapid increase in processing volumes since the middle of the last century. Today, plastics are indispensable in virtually all industries and are an essential part of our lives.

For some years now, however, we have been confronted with numerous images and reports about illegally operated landfills in Asia and Africa, about the disposal of plastic waste into rivers, which then flows into the world's oceans, about the consequences of plastics in the oceans with regard to animals and plants, about health risks for humans due to improper incineration of plastic waste, and about the social, economic, and political challenges that arise as a result. The problems are fundamentally attributable to the improper handling of waste and the sometimes excessive use of short-lived plastic products. One consequence of this development, however, is that plastics as a material are increasingly subject to blanket rejection, and a factual discussion is often absent against the backdrop of striking images of animals dying from plastic waste.

In numerous applications, plastics are also the ideally suited material from an ecological perspective. Furthermore, the group of thermoplastics in particular is fundamentally recyclable due to the possibility of remelting.

In this chapter, the critical aspects of the use of plastics and the handling of plastic waste are first examined. A historical classification and the current state of production, processing, collection, and recycling of plastics is the subject of a further section (Section 1.2). This is followed by a discussion of the material plastic as such, and the property profile, production, and recycling options of the most important plastics by volume are presented. Furthermore, an analysis relating to Germany is carried out of the prerequisites that must be in place in order to recover the valuable material plastic from plastic waste.

1.1 Plastics Under Criticism – Environmental Issues

The handling of plastic waste is organized very differently in the countries of the world. While many countries have a functioning waste management system that ensures the orderly collection of waste, there are others in which collection is not organized or is only inadequately organized. The result is the improper disposal of a large proportion of municipal solid waste, including plastics. But even in countries with a reasonably functioning waste system, waste is frequently discarded carelessly into the environment. Reasons for this range from insufficient sense of responsibility, convenience, indifference, and ignorance to the expression of provocation by people against the general order.

Without wishing to downplay the problem, the criticism of the waste that is disposed of uncontrolled in the environment overall seems to focus disproportionately on the quantitatively rather small share of plastics. The following reasons may have contributed to this phenomenon:

- Plastics originate from the chemical industry and are therefore frequently regarded by the public as harmful to health.
- Plastics have a long service life and therefore persist in the environment for a long time.
- Plastics have a low density and are therefore easily distributed in the environment, for example from open landfills by wind. In addition, plastic parts float on the water surface due to the difference in density and are therefore more visible than waste made of metals.

That the lack of waste management systems in some countries of the world has led to major environmental problems is evident. In addition, there is a lack of fact-based public education about the valuable material plastic in many respects and its advantages in everyday life, as well as about the responsible handling of plastic waste. Furthermore, the topic of microplastics should be brought more into focus, since retention measures can do a great deal to reduce the input of microplastics into the environment. So far, too little is known about the regionally very different mechanisms that lead to the formation and spread of microplastics.

In the following, some of the resulting environmental problems are presented on the basis of studies.

1.1.1 Plastics in the Oceans

There are numerous studies that estimate the occurrence of plastic waste in the world's oceans and the annual input of plastic waste into the world's oceans using different modeling approaches. It is clear that all studies involve estimates and extrapolations, because the quantities of waste can only be determined with great effort. The modeling approaches are entirely different and are based either on plastics drifting on the surface of the seas determined by sampling, or on uncontrolled quantities of waste on land that can potentially reach the seas. Only a rough picture of the situation can therefore be outlined.

Annual Input of Plastic Waste into the Oceans

plastic waste is transported into the world's oceans either via land or reaches them at sea through accidents or fishing. According to a study by McKinsey, the quantity of waste released at sea accounts for less than 20% of the total quantity of waste entering the oceans annually [NN01-ol, p. 7]. This figure is also confirmed by a study by Morales-Caselles et al., which assumes a share of 22% of the waste in the sea that enters the oceans through maritime activities (fishing, etc.) [MVG+21, p. 2]. Thus, the largest proportion of waste is transported from land into the seas, even though the share of 20% of waste released at sea is certainly not negligible.

Jambeck et al. analyze the quantity of plastic waste occurring in coastal regions worldwide (50 km distance from the sea) that is either collected in an uncontrolled manner and enters the environment directly or is stored on open landfills. The approach of the study assumes that this waste can potentially reach the sea. Taking into account the population living in the regions studied, a total quantity of plastic waste of 99.5 million tonnes in 2010 is calculated, of which 31.9 million tonnes are classified as the proportion that is not treated by a controlled waste management system. Assuming a rate of 15 to 40% of this proportion entering the seas, waste quantities of 4.8 to 12.7 million tonnes can be calculated that are introduced into the oceans worldwide every year. Approximately 55% of this waste is caused by China, Indonesia, the Philippines, Vietnam, and Sri Lanka [JGW+15]. In a further study, Lebreton et al. examine the worldwide input of plastics into the seas via rivers. According to this, 1.15 to 2.41 million tonnes of plastics enter the world's oceans via rivers every year [LZD+17]. Schmidt et al. also show that the input of plastics via rivers accounts for a substantial proportion of the total quantity of plastics transported from land into the seas [SKW17]. From a large number of recognized studies, the scientists extracted a dataset and extrapolated the input of plastics via the catchment areas of rivers into the seas using a modeling approach. An additional decisive factor is the freshwater input of individual rivers into the oceans worldwide, which is taken into account using the database of Milliman and Farnsworth [MF11]. The basis for the underlying waste quantities is the quantity of waste calculated in [JGW+15] that is not captured by a controlled collection system, supplemented by data from a further study by Hoornweg and Blade-Tata [HB12], in order to include not only the coastal areas but also the catchment areas of the rivers. Depending on the respective database regarding the concentrations determined by sampling in rivers, input quantities of 0.48 to 2.75 million tonnes of plastics via river systems into the seas are calculated worldwide. Approximately 90% of this quantity enters the seas via ten rivers of the world, of which eight are located in Asia and the other two in Africa.

With regard to Europe, González-Fernández et al. estimate in a study published in June 2021 that between 307 million and 925 million pieces of waste are transported into the seas via river systems in Europe. The plastic fraction is 80% and consists essentially of packaging, bottles, and bags. In total, according to this study, between 1,600 and 5,000 tonnes of plastic waste per year enter the seas from Europe. Turkey, Italy, the United Kingdom, and Spain together account for the largest share of the waste, with 45% of the total quantity. According to this, only approximately 2% of the European total quantity enters the sea from Germany's river systems, which corresponds to a waste quantity of 32 to 100 tonnes per year [GCH+21].

Worldwide Occurrence of Plastic Waste in the Oceans

The plastic waste introduced from land into the seas accumulates partly in coastal waters, bays, and gulfs, and partly in large zones within the five ocean gyres (North Pacific Gyre, North Atlantic Gyre, South Pacific Gyre, South Atlantic Gyre, and Indian Ocean Gyre) [ELC+14]. The largest accumulation of waste is located in the North Pacific Gyre, which bears the name Great Pacific Garbage Patch and covers an area of 1.6 km^2. According to Lebreton et al., between 45,000 and 129,000 tonnes of plastic waste are present in the near-surface area of this zone. This consists mainly of fishing nets and other fishing material, as well as bottles and their caps [LSF+18]. Further studies deal with the plastic waste distributed worldwide in the seas and calculate the quantity of waste per km^2 of ocean surface area using counts of numerous samples, the underlying ocean currents, and other factors [e.g. ELS+14, SWL+14]. This results, for example, in world maps with model-derived quantities of waste, as can be seen in Figure 1.1. Particularly for the macro fraction >200 mm, the five zones can be identified that are found in the five ocean gyres, as well as areas with high concentrations, for example in the Mediterranean Sea or the Gulf of Mexico.

Eriksen et al. and Cózar et al. compare in their studies on the amount of plastic waste present at the surface the calculated values with studies on the annual input quantities into the oceans via rivers and coastal areas and the estimated amount of waste present in the oceans. They arrive at values that are in some cases significantly lower and attribute this to the sinking of waste through the attachment of algae and microorganisms, through the degradation of plastic into the smallest particles (nanoparticles), or through organisms ingesting the waste fragments [ELC+14, CEG+14].

Through sampling in river catchments and the oceans, further insights into the waste present worldwide in the oceans can be derived. In particular, the type of waste item found can be used to draw conclusions about its origin and material. Morales-Caselles et al. classify in their study waste items documented in several databases across seven marine regions. In order to create comparable conditions, they first carry out a harmonized classification of items across all evaluated studies according to product type, material, origin, and region. Of a total of 12 million examined waste fragments, on average 80 % consist of plastic, followed by metal, glass, and fabrics (textiles). With 95 %, the largest proportion of plastic is found, as expected, at the ocean surface, since the density of many plastics is lower than that of water [MVG+21].

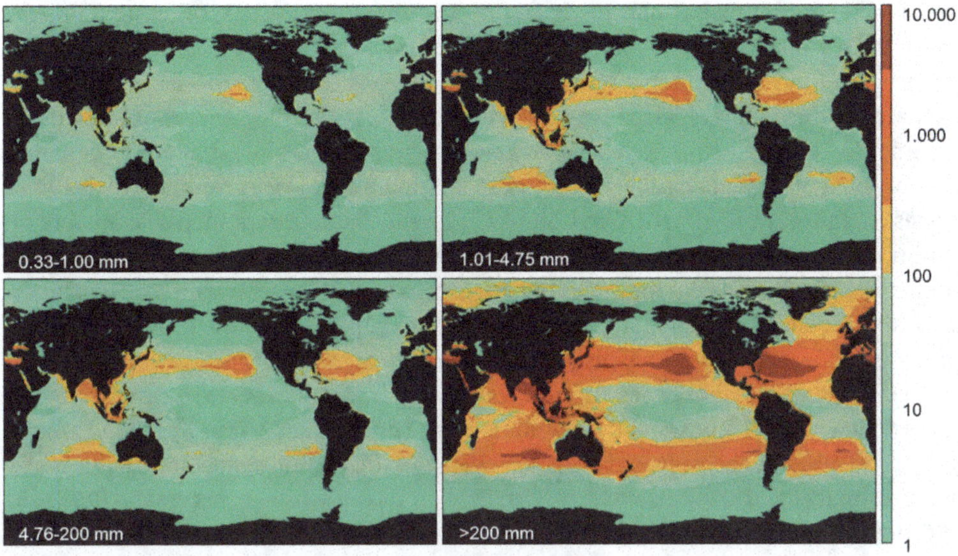

Figure 1.1 Model results on plastic waste occurrences in the oceans in g/km^2 for four different size classes of floating waste pieces present [ELC+14]

Figure 1.2 to Figure 1.4 show a summary overview of waste items found in rivers, in coastal waters, and on the open sea, each at the water surface and on the bottom (riverbed or seabed). It can be seen that at the water surface in rivers and at the coast, plastic waste primarily occurs in the form of plastic bags, food containers, and plastic bottles (approx. 65 % by number of items). Plastic bottle caps account for as much as 5 % of the total quantity. On the open sea, approximately 20 % plastic bottles and bottle caps are still found; however, waste associated with maritime activities accounts for the largest share at approx. 60 % (ropes, fishing accessories, fishing nets). It is notable that on the open sea, film-like packaging such as bags or pouches occurs only in a very small proportion. Since these mostly consist of lightweight polyethylene types, it would seem plausible that they are also transported to the open sea via wind and ocean currents. According to the cited source, however, this is not the case, as is also confirmed in [LSF+18]. This shows that there is evidently no strong exchange between waste in coastal waters, which is introduced from land, and the open sea. At least near the surface, waste from rivers is retained in the coastal zone and only reaches the large ocean gyres at specific locations via corresponding ocean currents [MVG+21].

By comparing the recorded waste items at the surface and on the bottom of rivers,, coastal waters, and the open sea, it becomes clear that a considerable proportion of plastic items is also found on the bottom of these marine regions. While plastic bags in rivers are still largely found in the surface water, the proportion of bags in the waste on the bottom of coastal waters and the open sea increases to as much as 18.5 %. It is also notable that plastic bottles represent the largest proportion of waste items on the seabed in open waters, while items from fishing activities tend to be found at the surface. Whether plastic items remain near the surface or sink to the seabed depends on many factors. On large-area waste items such as bags or films, microorganisms or algae settle, causing a shift in density. In general, the buoyancy of waste items depends not only on density but also on the displaced volume, so that large-volume waste items possess high buoyancy, particularly empty and sealed containers such as canisters or bottles. Waste with high buoyancy is primarily washed onto coasts and beaches by wind and near-surface currents. On the other hand, plastic waste with low buoyancy tends to sink and is driven along coasts or out to the open sea by the prevailing ocean currents. With regard to the vertical and horizontal transport of plastic waste in the sea and the depth-dependent concentrations calculated from this, several differing studies exist, which collectively show that, among other things, the current and wind conditions prevailing in the respective sea areas have a major influence and that a more detailed analysis of the particle size distribution of the waste is necessary [RSN+15, CEG+14, LMG+14].

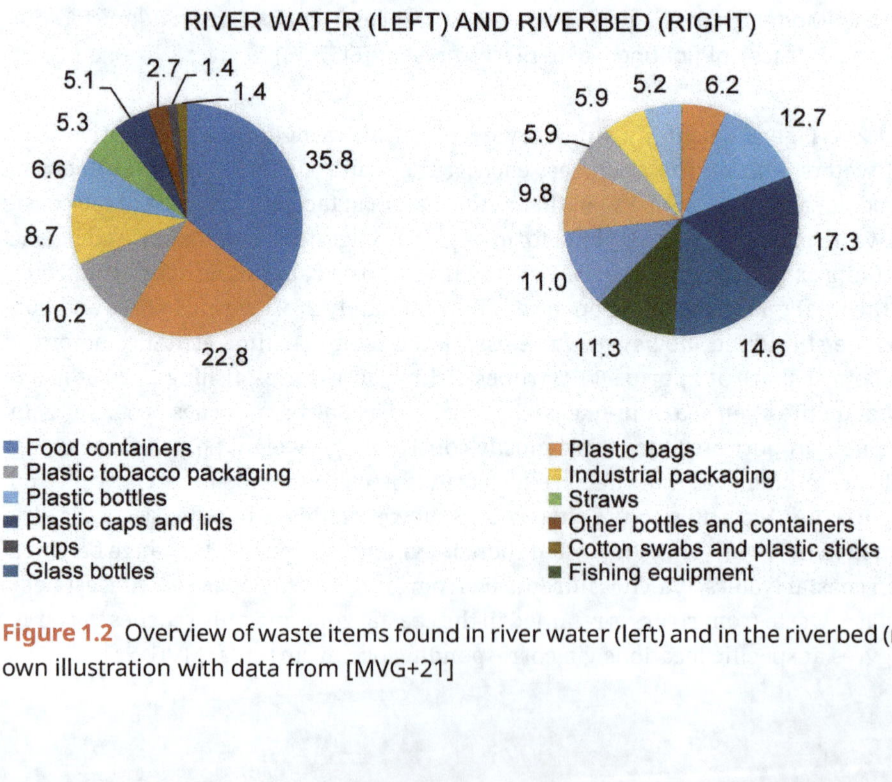

Figure 1.2 Overview of waste items found in river water (left) and in the riverbed (right), own illustration with data from [MVG+21]

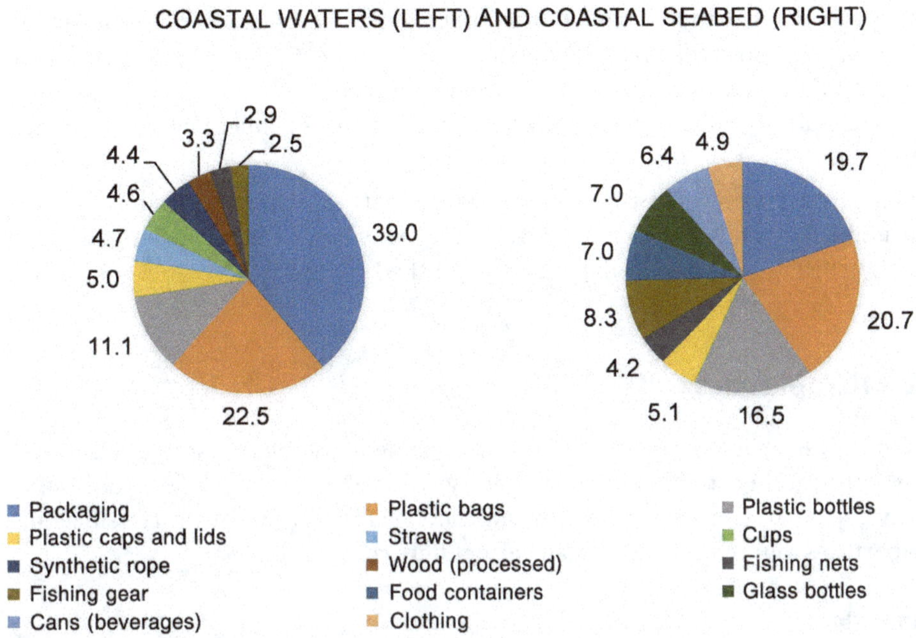

Figure 1.3 Overview of waste items found in coastal waters (left) and on the coastal seabed (right), own illustration with data from [MVG+21]

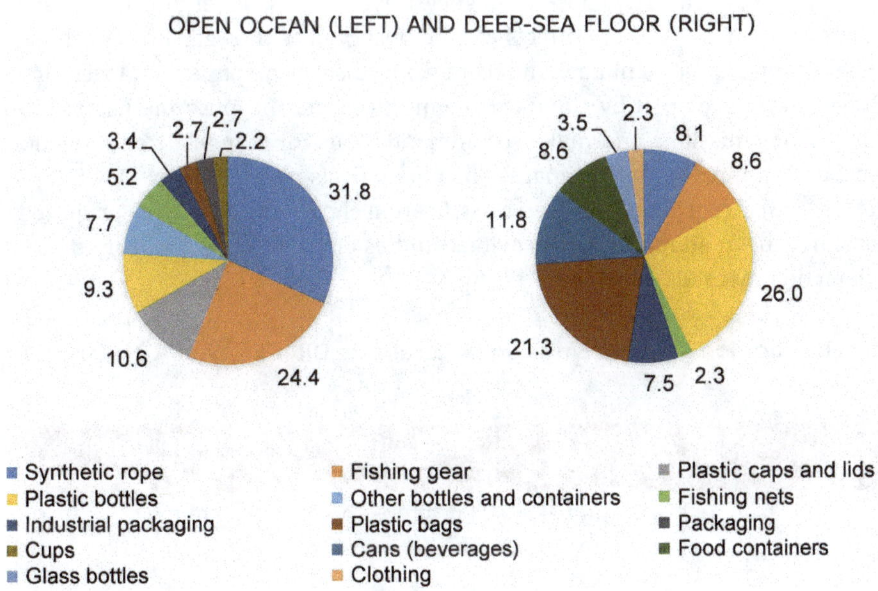

Figure 1.4 Overview of waste items found on the open sea (left) and on the deep-sea floor (right), own illustration with data from [MVG+21]

The studies on environmental pollution by macro- and microplastics, published in increasing numbers in recent years, show that a unification of different study results is extremely difficult, because the methods are not uniform and because different units are used to quantify marine pollution, which in some cases cannot be converted into one another.

Overall, the studies present extrapolated model results for which partly unavoidably vague assumptions must be made. A worldwide analysis thus exhibits a high degree of scatter, which may be reduced by newly published studies.

1.1.2 Microplastics

On the occurrence of microplastics in the environment, there are numerous publications whose differing investigation methodologies have led to a wide scatter of results. Not only is sample collection in water or soil handled differently, but the classification of the particles found by size and shape also differs considerably.

Classification

Since the negative effects of plastic emissions on the environment and humans are still insufficiently described, a classification by size and shape depending on the hazard potential cannot be made. In this respect, the seemingly existing material group "microplastics" is an arbitrarily chosen category. Nevertheless, depending on the size of the plastic emissions, different effects can be observed with regard to origin, occurrence, and environmental impact. In recent years, discussion papers and technical reports have been developed by standards committees that propose the use of size classes for plastic emissions and make recommendations for standardized analytics with regard to sampling, sample preparation, and analysis methods [DIN CEN ISO/TR 21960 and BJG+18]. Accordingly, the size classification shown in Table 1.1 is proposed, which has partly been adopted from previous studies and partly results from the respective detection methods to be used.

Table 1.1 Classification of Plastic Emissions According to DIN CEN ISO/TR 21960:2021-02

Nanoplastics	Microplastics	Large microplastics	Macroplastics
<1 µm	1 µm–1000 µm	1–5 mm	>5 mm

Classification is usually based on the greatest length of the particles, whereby films and fibers with large dimensional differences are sometimes classified in a different manner [BBB+21, p. 20; BDS+22, p. 14].

With regard to the formation of microplastics, a distinction is made between primary and secondary microplastics. Primary microplastics enter the environment directly in the form of small particles, while secondary microplastics are introduced into the environment as macroplastics, which then break down into fragments through environmental influences. Primary microplastics can be further subdivided as follows [BBH+18, p. 9; BBB+21, p. 16/17]:

- Type A: deliberately produced primary microplastics, e.g. abrasives in cosmetic products, polymeric blasting media, laser sintering powder
- Type B: microplastics generated during the use phase of a plastic component through mechanical stress or environmental influences, e.g. tire abrasion, weathering of paint, textile fibers released during washing

Entry into the Environment – Quantities and Proportions

According to Bertling et al., Type B accounts for 89% of the total quantity of primary microplastics in Germany, with tire abrasion representing the largest share. Overall, according to the authors' estimates, plastic emissions in Germany consist of only 26% macroplastics and 74% microplastics [BBH+18, p. 15]. This means that in Germany, compared to the global perspective, the microplastics issue is more important than the introduction of macroplastics into the environment when it comes to avoiding emissions (see also [BDS+22, p. 17]). This finding contrasts with the consideration of global plastic occurrences in the world's oceans. Figure 1.5 shows the proportions of quantities of microplastics and macroplastics occurring in the world's oceans depending on their respective origin. Thus, the proportion of macroplastics in the ocean is estimated at 92%, of which 78% originates from land-based sources. Accordingly, only 8% of all plastic waste in the ocean can be attributed to microplastics, three quarters of which come from land-based sources. It should be noted as a limitation that a not insignificant portion of microplastics never reaches the ocean at all, but instead remains on land, for example through sedimentation. Thus, depending on the study, the proportion of microplastics that does not reach the oceans is estimated at 53 to 98% [BBH+18, p. 25; GKR20, p. 37; She16].

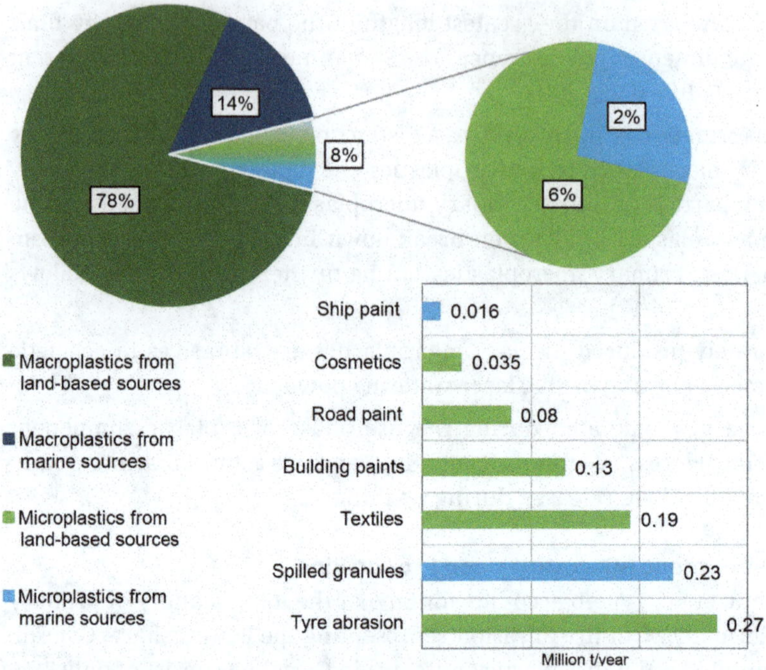

Figure 1.5 Overview of the proportions of macro- and microplastics in the world's oceans, broken down by origin, own illustration based on data from [GKR20, p. 37] and [She16-ol]

Origin of Microplastics

In many publications, analyses and extrapolations are used to provide information on the origin of the microplastics found. Thus, most studies identify tire abrasion (natural rubber and synthetic rubber) as the largest source of microplastics by quantity [BBH+18, p. 10; BBB+21, p. 26; BDS+22, p. 32, BF17, p. 21; She16]. Textiles are widely considered another major source of microplastics [BF17, p. 21; NT16]. Breitling et al. identified a total of 74 potentially relevant sources, of which 51 were also quantified using estimates. Accordingly, the ten main sources of microplastics are [BBH+18, p. 10/11]:

1. Tire abrasion
2. Release at construction sites
3. Release of microplastics during waste disposal
4. Abrasion of shoe soles
5. Abrasion of polymers and bitumen in asphalt
6. Abrasion of plastic packaging
7. Granule loss

8. Abrasion of road markings
9. Wind dispersal on playgrounds and sports facilities
10. Fiber abrasion from textile washing

Transport Pathways and Fate of Microplastics

The microplastics introduced into the environment from various sources either remain at the site of the original entry or are subject to transport on land or ultimately to water bodies or the oceans. The recovery of microplastics from the environment is also in principle worth considering. However, this proves to be extremely difficult, as the low weight of the particles leads to wind dispersal or washout by precipitation. So far, in contrast to the recovery of macroplastics by municipal waste management companies, there are hardly any analyses that quantify the amount of recovered material [BBH+18, p. 19].

According to [BDS+22], the following transport pathways for microplastics can be distinguished:

- Point-source emissions
- Diffuse emissions
- Treated wastewater from sewage treatment plants
- Transport by wind
- Stormwater drains in separate sewer systems
- Runoff by rainwater
- Combined sewer overflows
- Groundwater
- Directly discharging households/industrial facilities
- Transport by animals
- Stormwater drainage outside of settlements
- Waves
- Direct discharges from ships and hydraulic structures

Fibrous microplastic particles can be transported by wind over long distances through the air. These airborne particles, originating mainly from textiles, are present in indoor and outdoor air or in settling dust or in substances accumulating on the ground through precipitation. Studies have shown that approximately 30 % of fibrous particles are of petrochemical origin [DGM+17; DGS+16, WL20]. The transport of microplastics by wind is strongly dependent on the region, and it has been shown that, for example, in mountain regions the distribution of microplastics appears to be regionally limited, and additionally transport via precipitation is more significant due to large differences in altitude than in flatter regions [AAP+19, BDS+22, p. 18]. The transport behavior of microplastics by wind and in aqueous environments depends on the density of the plastic type (in relation to water), the size of the particles, the external convective forces, and the substrate. Polyolefins and foamed plastics can easily be transported over long distances, while engineering plastics or elastomers tend more toward sedimentation [BBH+18, p. 20].

Another significant transport pathway for microplastics is determined by the water management systems of cities and villages, which affect both stormwater and wastewater. In Germany, a distinction is made between the combined sewer system, in which both types of wastewater are directed to the sewage treatment plant, and the separate sewer system, in which stormwater is not treated. The combined sewer system therefore operates more efficiently with regard to the retention of plastic emissions than the separate sewer system. In sewage treatment plants, microplastic particles are separated to a high degree (mass-based separation rate of 95 % according to [BBH18, p. 23] or 99 % according to [WL20, p. 229]). The stormwater accumulating in the separate sewer system that is not treated, as well as water not discharged via the sewer system, is not treated. However, it should be mentioned that outside of sewage treatment plants there are additional sinks for microplastics. These include, for example, the sediment traps in gully pots, stormwater retention basins, or infiltration systems [BBH+18, p. 21].

Microplastics that cannot be retained enter the environment and accumulate in soil, rivers [HSW+17], lakes [VTR17; THR+19], on the seabed [LSL+17] or even in sea ice [PPB+19], or are washed up in coastal regions. Schernewski et al. analyzed the input of microplastics from land into the Baltic Sea and found that a large proportion of the microplastics introduced remains close to the point of entry at the coast and is not distributed further into the sea. According to this study, the residence time of microplastics in the open waters of the Baltic Sea is low at 14 days or a few months, contrary to previous studies. The input into the seas results mainly from overflow systems of sewage installations, which come into play particularly during heavy rainfall events [SRH+20; SRH+21].

However, microplastics also remain on land via various pathways and can be detected through soil investigations (e.g. [LLS+18]). According to Welden et al., the application of sewage sludge is the main source of microplastic input onto land [WL20, p. 229]. In Germany, due to the associated soil contamination, the application of sewage sludge is limited under § 14 of the Sewage Sludge Ordinance to a maximum of five tonnes of dried sewage sludge per hectare within three years. Nevertheless, approximately 35 % of sewage sludge, potentially containing microplastics, re-enters the environment through agriculture and landscaping [BBH+18, p. 24].

Effects of Microplastics on the Environment

The effects that microplastics exert on the environment have not yet been sufficiently clarified. Due to their size, the ingestion of microplastics by animals is a major problem, particularly for marine organisms. This can lead to blocked digestive tracts and ultimately to the death of the animals, which has been documented in numerous cases [Ker16].

Since plastics are generally inert and are considered toxicologically safe, they do not pose a major hazard from a chemical-toxicological perspective. However, it must be taken into account that additives or degradation products can be released through environmental influences. To date, hardly any studies are known regarding microplastics and the potential for chemical harm to animals and humans [BBH+18, p. 31]. Approaches to the damaging influence on ecosystems, particularly for nanoparticles, are discussed in [MKZ+17], [RH18] or [SWL+18].

1.2 Historical Development and Current Status

The following provides a brief global overview of the development of plastics, production volumes, and the treatment of plastic waste.

The history of plastics can be traced far back into prehistoric times, when natural polymers such as wild rubber, natural resins, or plastics made from protein substances were processed into moldable compounds through chemical changes that were unknown at the time, under the influence of temperature. By adding color pigments or other fillers, the properties of these plastic molding compounds were altered, just as is still done today through the use of additives.

From the 19th century onwards, various natural materials such as horn or natural resins were increasingly used as substitutes for expensive natural materials such as marble or ivory. In parallel, the history of rubbers developed with significant milestones such as the discovery of mastication of rubbers to improve processability in 1824 by the Englishman Thomas Hancock, or the discovery of vulcanization in 1839 by Charles Goodyear.

At the beginning of the 20th century, the first synthetic plastics emerged. The term "Kunststoff" (plastic) was coined in the German-speaking world during this period by the founder of the journal "Kunststoffe", Dr. Ernst Richard Escales. In addition to phenolic resins, significantly influenced by the inventor of the material "Bakelite", Dr. Leo Baekeland, polyolefins, polystyrene, and polyvinyl chloride were developed and industrially produced from the beginning of the 1930s [Bra13]. From this time onwards, a rapid development of plastics began, which is illustrated in terms of production volume in Figure 1.6 over time.

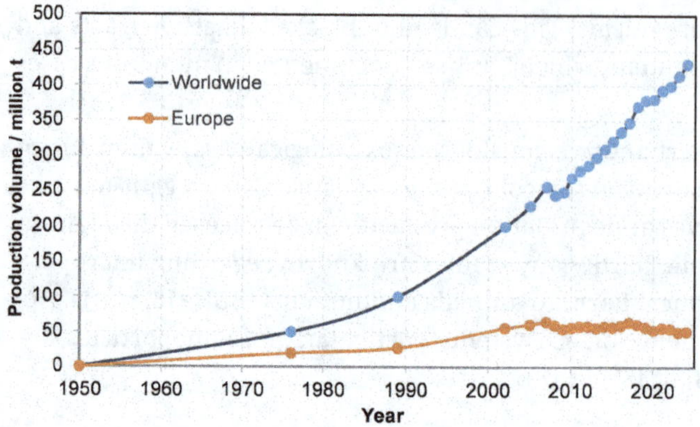

Figure 1.6 Development of plastic production volumes in Europe and worldwide, own illustration based on data from Plastics Europe

Global production volumes have been rising steadily, particularly since the 1970s, and reached a value of approximately 431 million tonnes in 2024. Values for Europe stagnated between 2007 and 2021 at a level of between 57 and 64 million tonnes [NN02-ol]. Since 2023, production volumes have been declining annually, which represents a significant reduction in the share of European production in global output. In Germany, production volume in 2023 was approximately 10.5 million tonnes, corresponding to a share of just over 20 % of European production capacity, which is also declining compared to previous years. Demand for plastic as a raw material by plastics processors amounted to 50.3 million tonnes in 2021, of which approximately one quarter was attributable to Germany [NN02-ol, p. 31].

The plastics most widely used across Europe in 2021 are the polyolefins polypropylene (PP) and polyethylene (PE) with its variants HDPE (high-density PE) and LDPE (low-density PE). These account for approximately half of the total volume of plastics processed in Europe. As can be seen from Figure 1.7, polyvinyl chloride (PVC), polyurethane (PUR), polyethylene terephthalate (PET), and polystyrene (PS) are further plastics of significant importance in terms of volume.

The most important application area in terms of volume of plastics, packaging accounts for nearly 40% of the plastics processed in Europe. These include food packaging, logistics and transport packaging, packaging for medical purposes, garbage bags, and household and food storage bags. The most commonly found plastics in this segment are PE, PP, and PET. As is clearly shown in Figure 1.8, the construction industry uses a further 21.3% of the processed plastics volume.

Products manufactured include those for road and civil engineering (e.g., PVC pipes), for structural engineering, for numerous building materials, for sanitary applications, floor coverings, wall and ceiling claddings, window profiles, and insulation materials.

The most important plastic type for these applications is PVC, followed by PE, PP, EPS (expanded polystyrene), and PUR.

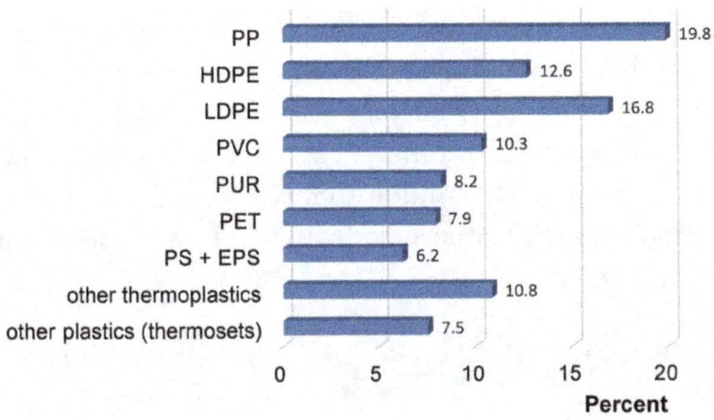

Figure 1.7 Demand for individual plastic types by European plastics processors in 2021, own illustration with data from Plastics Europe Market Research Group (PEMRG) and Conversio Market & Strategy GmbH

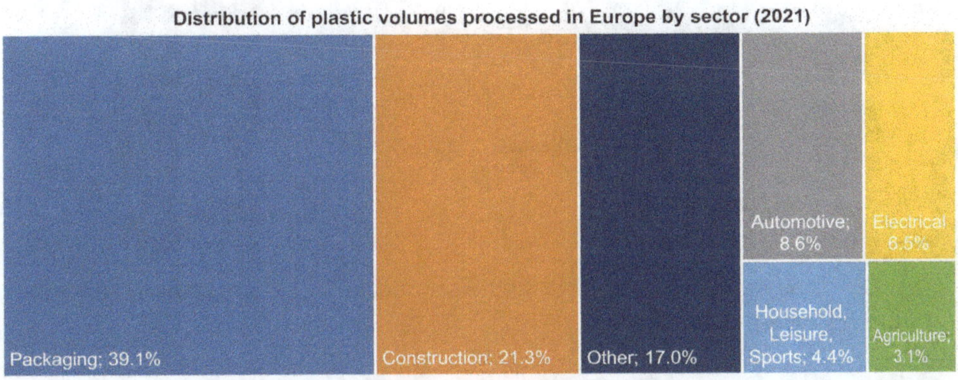

Figure 1.8 Breakdown of the plastics volume processed in Europe by industry sector (EU28 + N/CH), own illustration with data from Plastics Europe Market Research Group (PEMRG) and Conversio Market & Strategy GmbH

The automotive industry accounts for a further nearly 9% of the total plastics volume. In addition to PP, engineering plastics such as PC/ABS, PA, PUR, or PC are increasingly used for the manufacture of numerous interior and exterior parts. Other notable sectors include the electrical industry, the household, leisure and sports sector, and agriculture.

A more detailed examination of the plastic types, their market significance, recycling options, applications, and their use in selected material streams in Germany is provided in Section 1.3.1 and 2.4.

A worldwide analysis of total plastics production and plastics waste volumes within a single year is the subject of a 2016 study by McKinsey & Company, Inc. According to this study, of the 330 million tonnes of plastics produced, 150 million tonnes were used for short-lived applications that become waste within the year of production, such as food packaging. The remaining 180 million tonnes were processed into long-lived products which, by the authors' definition, only become waste in subsequent years, such as products from the construction sector. From the category of long-lived applications (manufactured in previous years), 110 million tonnes became plastic waste in 2016, which, combined with short-lived products, adds up to a total of 260 million tonnes of plastic waste worldwide (see Figure 1.9).

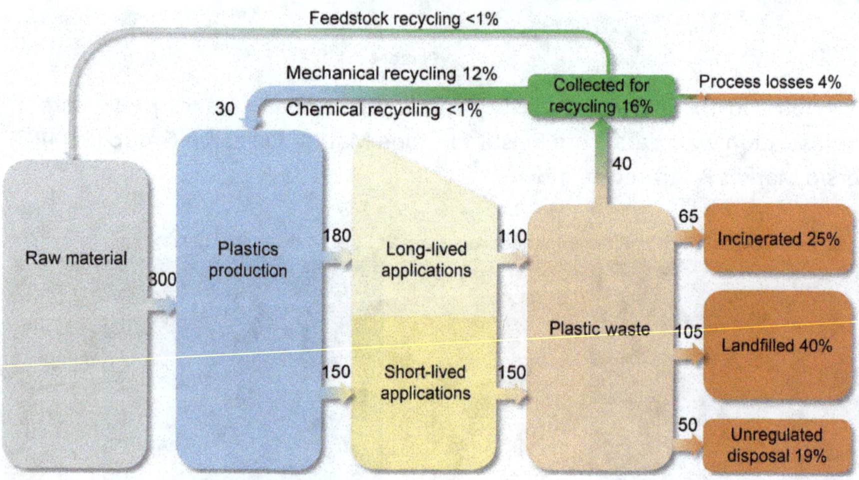

Figure 1.9 Global plastics and plastic waste production in 2016 in million tonnes, according to [Mck16-ol]

As explained in Section 1.1.1, in many countries a large proportion of waste is not collected in a controlled manner and therefore not properly disposed of. According to this study, this accounts for 19% of the total plastic waste volume. A further 40% is landfilled, 25% is recovered for energy, and only 16% is sent for recycling.

With regard to long-lived durable goods, the study allows the conclusion that more plastic products are manufactured each year than the amount of plastic waste generated in the same year. If this trend continues, the entire market could not be supplied with secondary plastics even at a theoretical recycling rate of 100%. Fossil raw materials would still be necessary, unless plastic waste landfilled in previous years could be used as a raw material source or the extraction from renewable raw materials were to gain substantially in importance. This is at least not to be expected for engineering plastics, which are frequently used for long-lived products. McKinsey assumes that the volume of plastics processed will increase by nearly 70% to 560 million tonnes by

2030, and that the waste volume will also rise by 70% to 440 million tonnes. At the same time, however, this forecast predicts that the recycling share will increase from 16 to 50% and that unregulated waste disposal will fall to below 1% [Msk16-ol]. The latter, however, can only be achieved by establishing a functioning waste collection system in virtually all countries of the world. Lebreton et al. show in a model calculation that globally mismanaged plastic waste will increase linearly until 2060. Through improvements in waste management and an additional reduction in plastic consumption, this waste volume could be significantly reduced compared to the current level [LA19].

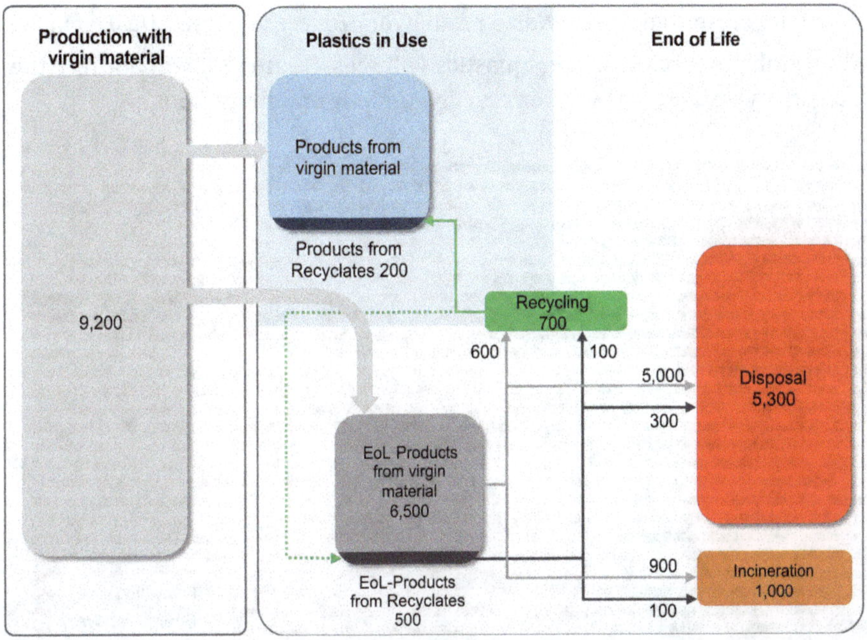

Figure 1.10 Cumulative overview of plastics produced worldwide from 1950 to 2017, those currently in use, and those treated differently after use, in million tonnes, own illustration based on and with data from [Gey20, p. 27]

In a widely cited study published in the journal Science in 2017 on determining the total amount of plastics produced since 1950 and analyzing their fate ('Production, use, and fate of all plastics ever made'), the authors show that nearly 80% of all plastic waste generated has either been landfilled or improperly disposed of in the environment [GJL17]. Figure 1.10 shows the relationships presented in the aforementioned study with more recent data published in 2020. In total, 9.2 billion tonnes of plastics have been produced worldwide, of which 2.7 billion tonnes are currently in use, incorporated into products. Over time, 7 billion tonnes of plastics have reached end-of-life (EoL) status and have been disposed of (75.7%), recovered for energy (14.3%), or recycled (10%). If the forecast from the McKinsey study is taken as a basis, the cumulative increase in the disposal category up to 2030 will be considerably smaller than that for incineration and recycling. A look at the European figures from 2019 shows that Europe contributes significantly to this development.

Recycling rates in all European countries are above 20% and the landfill rates in two thirds of the countries are below 40% (see Figure 1.11). Compared to 2006, the recycling rate in Europe has increased by 110% and the rate of energy recovery by 76%, while the landfill rate has decreased by 57% [NN02-ol, p. 54]. The positive development of rising recycling rates and increased resource efficiency in Europe is attributable to the EU regulations presented in 2.2 (Waste Framework Directive, Green Deal, etc.).

Following this global overview of total plastics volumes, Section 1.3 will address the individual plastic types and outline the recycling options and limitations.

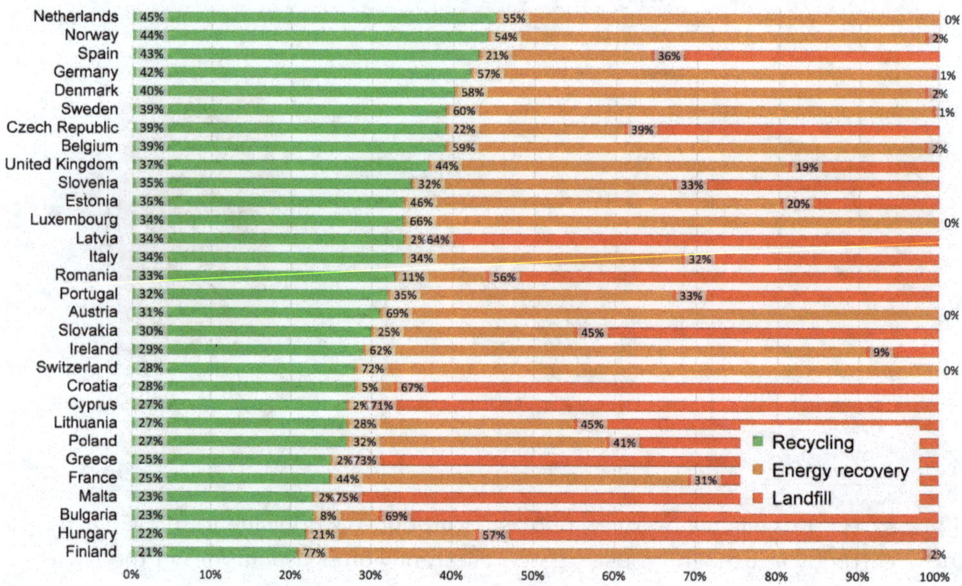

Figure 1.11 Shares of recovery methods for post-consumer waste in European countries in 2020 [NN02-ol, p. 49, slightly edited]

1.3 Plastics As a Material and a Valuable Resource

In the following, the plastics that are most significant in terms of volume for plastics processing are presented with regard to their property profiles, manufacturing methods, and recycling options. In addition, based on the fundamental prerequisites for obtaining secondary raw material plastics from waste, an overview of the complex interrelationships between legal requirements, the state of technology, and the economic viability of processing routes will be provided.

1.3.1 Property Profile and Classification of Plastics

Plastics offer numerous advantages over metals or ceramics. They exhibit very good weight-specific mechanical properties, they can be processed with comparatively low energy input, they offer a high degree of freedom in modification, and are (in the case of thermoplastics) fundamentally recyclable. Figure 1.12 shows an overview of the special properties of plastics.

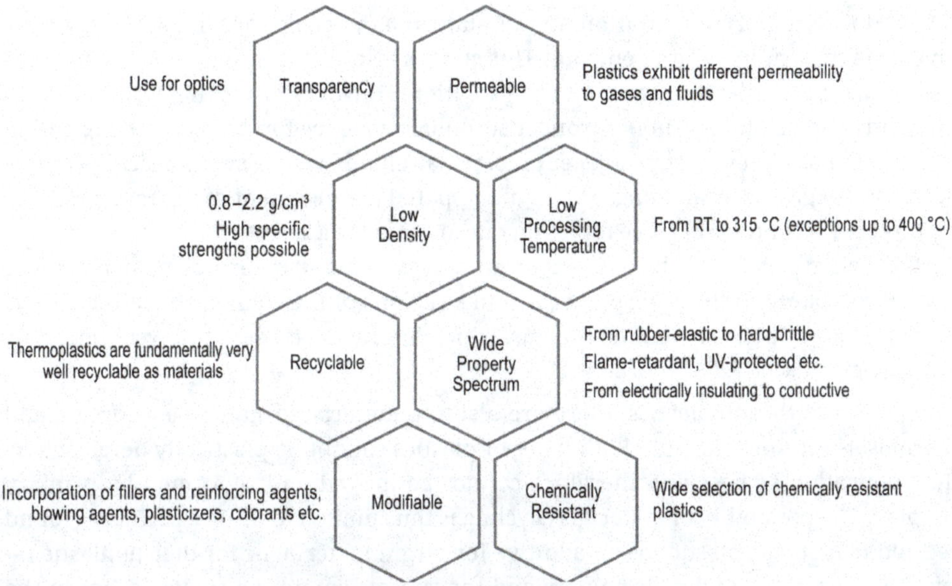

Figure 1.12 Selected special properties of plastics

Structure and Morphology of Plastics

Plastics consist of molecular chains, i.e., a sequence of monomers produced by a chemical synthesis reaction. The latter are the basic building blocks which, in the case of synthetically produced plastics, are obtained from crude oil by means of distillation and steam cracking. The monomers within the molecular chain are chemically bonded to one another via so-called covalent bonds. The actual cohesion of the polymer (cohesion) is, however, determined by the considerably weaker physical secondary valence forces acting between adjacent molecular chains. The magnitude of these binding forces directly influences the macroscopically resulting properties such as strength, chemical resistance, heat resistance, flow behavior, insulation properties, thermal conductivity, etc. [DHM22, pp. 73 ff.].

An increase in temperature leads to a reduction in secondary valence forces and thus to a decrease in strength. A further increase in temperature ultimately causes adjacent molecular chains to slide past one another when an external force is applied. The plastic is then in the melt state and can be processed using primary shaping processes (e.g., injection molding or extrusion). Secondary valence forces can also be reduced by solvents, and complete dissolution of plastics is possible when the appropriate solvent is used. Both melting and dissolving of plastics are essential for the material recycling of plastics. However, both are only possible for the group of thermoplastics. Within these, a further distinction is made between amorphous and semi-crystalline thermoplastics. Amorphous thermoplastics exhibit a random arrangement of molecular chains, also referred to as a coil configuration. In this state, the molecules are in an energetically favorable state of maximum entropy. This state is also found in semi-crystalline thermoplastics, particularly in the melt or in solution. However, when these cool down to the solid state, the increasingly stronger secondary valence forces lead to a disruption of this state of high entropy. Adjacent molecular chains with a symmetrical arrangement, low degree of branching, and regular spatial arrangement of the molecules exert directed secondary valence forces on one another. As a result, the molecular chains partially align parallel to one another, forming crystalline structures with high binding energy compared to the amorphous, coil-like regions of the chains. The latter always remain alongside the crystalline regions, which is why the term semi-crystalline thermoplastics is used [Kai16, pp. 84 ff.].

In addition to thermoplastics, there are also the material groups of elastomers and thermosets (or duroplastics). Both types of plastics cannot be plastically deformed by the application of heat and therefore cannot be melted and re-formed by primary shaping. The possibility of material recycling is thus limited to the use of finely ground regrind from these plastics as an additive for virgin material or for thermoplastic recyclates. The reason why elastomers and thermosets cannot be remelted lies in the cross-linking of adjacent molecular chains. Covalent bonds are formed by chemical reaction (in the case of elastomers also referred to as 'vulcanization'). Like the covalent primary valence bonds within the molecular chains, these cannot be broken by the application of heat in such a way that the chains slide past one another and the material flows. Elastomers have a low number of cross-linking sites, which, in combination with the low glass transition temperatures of rubber polymers, results in the typically rubber-elastic properties. Thermosets, on the other hand, are closely cross-linked after the curing reaction, which is reflected in the dimensional stability, strength, and hardness of the material [BBO+13, pp. 42 ff.].

For many years, in addition to thermoplastics, elastomers, and thermosets, the so-called thermoplastic elastomers (TPE) have also been partially classified as a separate material group. Depending on their composition, these are copolymers or polymer blends that exhibit thermoplastic processing properties (meltable) while possessing rubber-elastic service properties.

Modification of Plastics

In order to meet the requirement profiles of a wide variety of applications, plastics are frequently modified. Many properties can already be varied by changing the molar mass, the molecular shape, and the molecular order. Further possibilities arise from the production of copolymers or polymer blends as well as from the use of additives and processing aids.

Polymers with two or more different monomer types within a molecular chain are called copolymers. Like homopolymers, which contain only one monomer type, they are produced by chemical synthesis in reactors. Depending on the arrangement of the respective monomer types in the molecular chain, the following are formed:

- Alternating copolymers (monomers alternate in the chain)
- Statistical copolymers (random distribution of the monomers)
- Block copolymers (segments with several monomer units of one type alternate with those of the other type)
- Grafted copolymers (side chains of one monomer type are grafted onto the main chain consisting of the other monomer type)

A typical example is a copolymer of ethylene and propylene, which at a mixing ratio of 35/65 or 65/35 and by the addition of unsaturated hydrocarbons becomes an elastomer, EPDM (ethylene-propylene-diene rubber). Outside the mentioned mixing ratio, ethylene-propylene copolymers are used due to their high impact strength. Another industrially widespread copolymer is acrylonitrile-butadiene-styrene copolymer (ABS), which is also used because of its high impact strength, but also because of its higher strength, stiffness, and higher temperature resistance compared to polystyrene [DHM22, p. 95 ff.].

In contrast to copolymers, in which different monomers are linked together, the production of polymer blends involves mixing different, fully synthesized polymers with one another. This is done on typical compounding systems, such as a co-rotating twin-screw extruder. The properties of polymer blends are determined in particular by the intermolecular interactions of the two polymer phases. These depend on the compatibility of the two polymer types as well as the process parameters during the mixing operation. Even incompatible polymers can be blended using compatibilizers. Typical representatives of polymer blends are combinations of PC and ABS, PA 6 and PA 10, or LLDPE and LDPE. Since a large proportion of mixed plastics remains after plastic waste passes through a sorting facility, and the individual fractions are not entirely pure, blend technology also plays a significant role in the field of material recycling [BBO+13, p. 52 ff., DHM22, p. 102 ff.].

Another very frequently used possibility for modification consists in the incorporation of additives. This is done by a compounding process, where in most cases a co-rotating, closely intermeshing twin-screw extruder is also used. Plastics are often already equipped by the raw material manufacturer with additives such as stabilizers against UV radiation and oxidation or adhesion promoters. Compounds specifically tailored for a particular application are produced by compounders or raw material processors. However, some of the additives are only added to the plastic by the plastics processor.

In principle, two areas can be distinguished in terms of the objective of modification:

- Modification for the processing phase
- Modification for the service phase

A small selection of additives with their respective effects is shown in Figure 1.13. For processing, plastic melts with good flowability are advantageous, which can be achieved by lubricants. Nucleating agents are used to control crystallinity. Adhesion promoters form molecular bridges between fillers and the polymer matrix, thereby improving adhesion within the composite.

During processing, the polymers are exposed to high heat, and during the service phase to various environmental influences such as oxygen, ozone, UV light, etc. This can lead to chain scission and thus to degradation of the plastic. From a processing perspective, this results in a lower processing viscosity (higher melt flow index, MFI), and with regard to service properties, among other things, a loss of strength, stiffness, and impact strength. Stabilizers are intended to prevent thermomechanical, thermo-oxidative, and photo-oxidative degradation processes. Antioxidants are used against oxidative degradation, which is more pronounced at higher temperatures due to the higher diffusion rate of oxygen. Light stabilizers prevent yellowing, chalking, and changes in color and gloss. Heat stabilizers counteract thermal degradation during processing, which is particularly important for the production of PVC products. In particular, antioxidants, but also other stabilizers, are increasingly used in the recycling sector for re-stabilization. A problem here is that when a heterogeneous waste stream from unknown applications is used, a large number of different stabilizers are present. Re-stabilization can lead to unexpected effects, especially when partially degraded material is present in the material stream [BBO+13, p. 726 ff.; Kai16, p. 103 f.; Rei21].

For the use phase, a wide range of additives is available, among which reinforcing materials (e.g., glass, carbon, or aramid fibers) are of great importance, particularly for engineering plastics. Furthermore, cost-reducing fillers such as chalk are widely used. Flame retardants, antistatic agents, plasticizers, fungicides, and not least colorants are further additives that are used in appropriate concentrations depending on the application. The specific additivation of recyclates is always complicated by the fact that the composition of the incoming waste stream is frequently unknown and can only be determined with considerable effort.

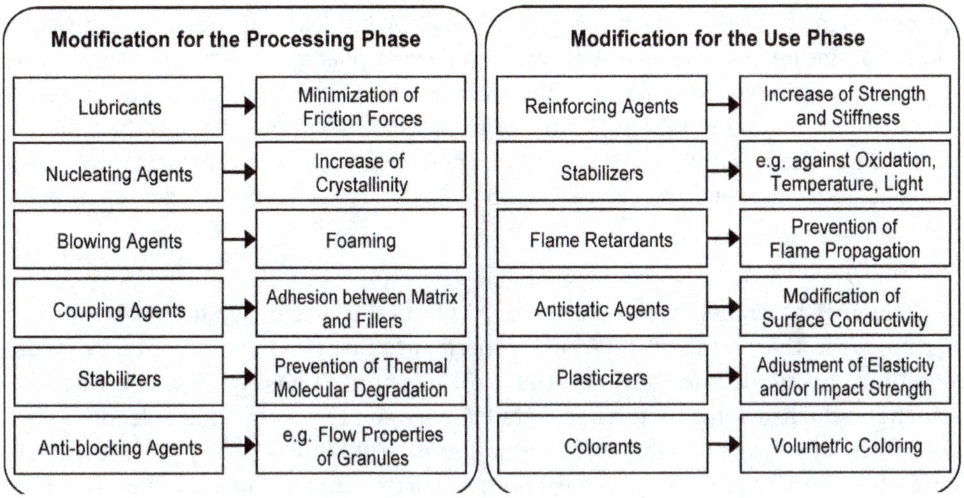

Figure 1.13 Selection of additives for the processing and service phases of plastics

Properties of Selected Plastics

In the following, the most commonly used plastics are presented in terms of their properties. A direct comparison of selected material characteristics is shown in Table 1.2. According to Figure 1.14, PP and the PE grades HDPE and LDPE are the most important plastics by volume. Even when looking at the waste fraction, it becomes clear that the quantity of these three plastic types accounts for the largest share at over 50%. As explained in Section 2.4, the material stream from packaging represents the largest share of plastic waste generated by volume. This also explains the high proportion of PP and PE shown here in the waste stream, since a large proportion of packaging is made from these plastics. PVC is also processed in a high proportion, which is largely attributable to the construction industry.

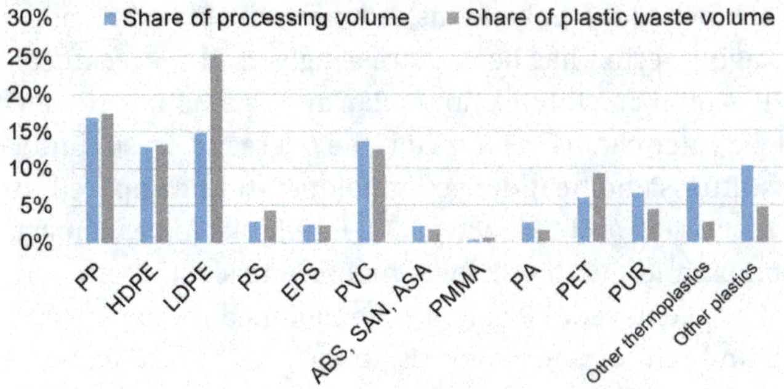

Figure 1.14 Shares of individual plastic types in total processing and plastic waste volumes in Germany in 2021, own illustration with data from [Con21]

PET occurs in the waste stream particularly as PET bottles, which explains the high proportion in the overall waste stream. Engineering plastics such as PU, PA, and the categories "other thermoplastics" and "other plastics" show lower shares in the waste stream compared to their shares in the total processing volume. The products made from them are frequently long-lived capital goods that may have only been developed and brought to market in recent years and have therefore not yet reached their end-of-life state.

Polyethylene (PE)

Polyethylene, which belongs to the standard plastics, is a semi-crystalline, non-polar thermoplastic that exhibits different property profiles depending on the reaction conditions during polymerization. LDPE (low-density PE) has short- and long-chain branches, which result in a greater distance between the main chains, leading to a low density and also a low degree of crystallinity. HDPE (high-density PE) and LLDPE (linear low-density PE), on the other hand, consist mainly of linear chains with few short-chain branches. With higher density and lower degree of branching, higher yield stresses tend to be achieved. In general, all PE types have very good electrical insulating properties, are chemically very resistant, and are very easy to process. Stiffness and strength, however, are rather low compared to most other plastics. With regard to permeability, PE has a relatively high gas permeability but a low permeability to water vapor, which is particularly relevant in the field of food packaging [Krü12, p. 432].

LDPE and LLDPE are used particularly in the film sector; HDPE is found in the area of blow-molded parts, pipes and profiles, and generally in the injection molding sector for products with rather low mechanical requirements. Thus, this plastic plays a major role essentially in the packaging industry, but due to its versatile adjustability also in the construction industry, the automotive industry, the electronics sector, and agriculture.

Polypropylene (PP)

Polypropylene, which like PE belongs to the polyolefins, is a versatile standard plastic that, in addition to the packaging sector, has been increasingly used for technical components for many years. PP is semi-crystalline, non-polar, and as a homopolymer (PP-H) has regularly arranged polymer chains. As a result, the degree of crystallinity is high, which leads to higher stiffness and heat deflection temperature compared to PE. The density is lower than that of PE, and below 0 °C, PP-H exhibits lower impact strength. As a copolymer in combination with ethylene, higher impact strength and higher transparency than PP-H are achieved. PP can also be modified in many ways by blending or by adding fillers and reinforcing materials. In the area of food packaging, PP is suitable for hot filling and can be steam sterilized [BBO+13, p. 460 ff.; Kai16, p. 269 ff.].

PP is used in the packaging sector, among other things, for liquid containers, films, and packaging straps; in the automotive sector for all types of injection-molded parts;

in the household sector for housings, film hinges, or containers, etc.; and in pipeline construction for hot water and wastewater pipes.

Polyethylene terephthalate (PET)

The plastic widely known for the PET bottle is a semi-crystalline thermoplastic that, depending on the type and processing conditions, can be present as an amorphous-transparent or semi-crystalline, opaque plastic. A distinction is made between crystalline PET-C, amorphous PET-A, and glycol-modified PET-G. The starting materials for the production of PET are ethylene glycol and terephthalic acid. For use as a PET bottle, amorphous PET is used, which can be set to be transparent with rapid cooling. Compared to PET-C, it has high impact strength, but slightly lower stiffness. The high barrier properties against CO_2 and O_2 qualify PET for use as a bottle for carbonated beverages. Compared to glass, PET is not breakable and has a significantly lower density. The higher stiffness and hardness of PET-C lead to the use of this type for technical parts. In addition, PET-C is abrasion-resistant and shows a low tendency to creep [BBO+13, p. 587 ff.; Kai16, p. 342 ff.].

In addition to its use as a bottle and tray for food, PET is used as an engineering plastic for electronic components, packaging straps, fibers, for many household items (e.g., microwave cookware), for films, or gears and sliding guides.

Polystyrene (PS)

Polystyrene, which belongs to the standard plastics, is an amorphous thermoplastic counted among the lower-cost plastics. As a homopolymer, it is a crystal-clear, brittle-hard plastic with extremely low impact strength and high susceptibility to stress cracking. There are numerous copolymers such as ABS, SAN, and ASA, which are superior to polystyrene in terms of impact strength, weather resistance, and heat deflection resistance. Polystyrene can be processed very well by injection molding or extrusion [BBO+13, p. 479 ff.; Kai16, p. 314 ff.].

Standard PS is frequently used in the packaging sector for, e.g., yogurt containers, clear packaging for toys, tools, etc., cups and bottles, and otherwise for electronic components, household articles, luminaire covers, glazings, rulers, set squares, or shower walls. PS foams form their own application group, with particle foams for rigid packaging parts, bicycle helmets, or impact sound insulation, or with extruded rigid foam boards (XPS) for insulation applications in the construction sector.

Polyvinyl chloride (PVC)

The vinyl polymers differ essentially in the variants without plasticizer (rigid PVC or PVC-U, unplasticized) and those with plasticizer (flexible PVC or PVC-P, plasticized). Both variants are produced via polymerization processes in which chlorine is added to ethylene. PVC is not processed in granule form like many other plastics; instead, the raw materials are compounded in powder form at the PVC-processing plant. Additives for processing and use are blended into the PVC raw material, and the mixture is then fed to the processing operation as a free-flowing dry blend. This is, on the one hand, more economical than prior granule production and, on the other hand, protects the residence-time-sensitive PVC material. Important additives are heat stabilizers, flow aids, and UV absorbers. Calcium carbonate is also frequently used as a filler to increase impact strength. PVC belongs to the engineering plastics and is a polar, largely amorphous thermoplastic with high strength and hardness as well as a high modulus of elasticity. The impact strength, particularly at low temperatures, and the barrier properties against gases such as N_2, CO_2, and O_2 are higher than those of polyolefins. By adding plasticizers, the secondary valence forces can be reduced, making the material more ductile. Flexible PVC is rubber- to leather-like, possesses very high impact strength and extensibility [BBO+13, p. 492 ff.; Kai16, p. 287 ff.].

PVC is used in the construction industry for pipes, window profiles, vinyl floor coverings, and generally for balls, sealing rings, electrical plugs, shoe soles, conveyor belts, coatings, films, and sheets, etc.

Table 1.2 Overview of Material Data, from [BBO+13; Kai16]

Plastic	Density [g/cm^3]	Melting temperature (TK) [°C]	Tensile modulus of elasticity [MPa]	Yield stress [MPa]	Heat deflection temperature HDT A 1.8 MPa [°C]
LDPE	0.915–0.92	105–118	200–400	8–10	-
HDPE	0.94–0.96	126–135	600–1400	18–30	38–50
LLDPE	0.935	126	300–700	20–30	ca. 40
PP-H	0.9–0.915	162–168	1300–1800	25–40	55–65
PET	1.38–1.4	250–260	2800–3100	60–80	65–75
PS	1.04–1.05		3100–3300	30–55	65–85
ABS	1.03–1.07		2200–3000	45–65	95–105
PVC-U	1.38–1.4		2700–3000	50–60	65–75
PC	1.2		2350–2400	60–68	120–130
PMMA	1.17–1.19		3100–3300	60–75	75–105
PA6	1.12–1.14	220–225	2600–3200 (dry) 750–1500 (cond.)	70–90 (dry) 30–60 (cond.)	55–80
PA66	1.13–1.15	255–260	2700–3300 (dry) 1300–2000 (cond.)	75–100 (dry) 50–70 (cond.)	70–100

1.3.2 Manufacturing and Processing of Plastics

In the following, the production of plastics starting from the basic raw material crude oil through to the polymer will first be outlined. In addition to synthetically produced plastics, there is also the possibility of obtaining plastics from plant-based raw materials, which are then marketed as "biopolymers". However, they have remained of minor importance to this day and are therefore not discussed here. Subsequently, the essential processing methods for plastics will be presented.

From Crude Oil to Polymer

The production of plastics was briefly outlined in Section 1.3.1 and will now be explained once more. For synthetically produced plastics, the basic raw material is crude oil, which is fractionated in a refinery by distillation. The share of the fraction used for plastics production accounts for only 4–6 % of total crude oil and natural gas consumption [Krü16, p. 4].

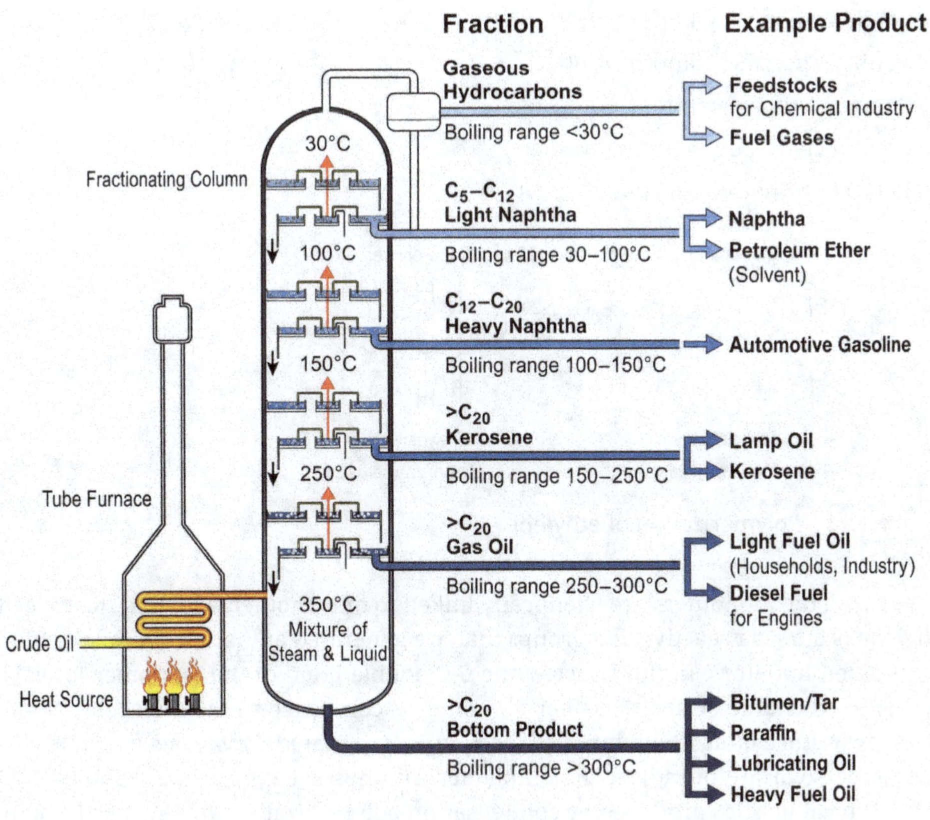

Figure 1.15 Distillation of crude oil, after [NN03-ol]

Figure 1.15 schematically shows the structure of a distillation plant. In the first step, the crude oil is heated to a temperature of approx. 350 °C, creating a vapor-liquid mixture that is fed into the fractionating tower. As can be seen in the figure, this tower consists of several levels, with the temperature decreasing from bottom to top. The vapor reaches the next higher level via so-called bubble caps. Depending on the boiling temperature of the hydrocarbon compounds present in the mixture, the respective fraction condenses on the corresponding level and can be drawn off there as a liquid phase. For further processing into monomers, the crude or light naphtha known as naphtha with C_5- to C_{12}-hydrocarbons and the liquefied petroleum gas are used. In a further process step, steam cracking, the long chains are broken down into shorter molecules at high temperatures. This produces, for example, ethylene, propylene, butylene and other compounds that serve as the basic materials for polymer synthesis.

The step from monomer to polymer takes place via so-called polymerization, whose process variants can be broken down as follows [Kai16, p. 41]:

- Step-growth reaction
- Condensation polymerization
- Addition polymerization
- Chain-growth reaction
- Radical chain polymerization
- Ionic/coordinative variants

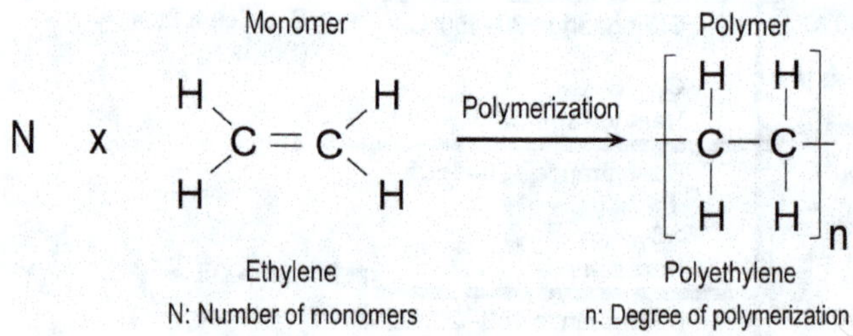

Figure 1.16 Polymerization of ethylene

In all cases, the monomers are chemically linked to one another, which in the case of ethylene occurs via reactive end groups that are generated at the start of the reaction by so-called initiators. In this process, the C-C double bond of the monomer loses its double bond character and a further monomer is connected to the molecular chain. In this way, the chains grow until the reaction is terminated by various mechanisms. This process variant of radical chain polymerization is used, for example, for PE, PP or PS. Typical plastics produced by condensation polymerization are PA6 or PET. Representatives of polyadducts (produced by addition polymerization as a step reaction) are PU or epoxy resin.

As can be seen from Figure 1.16, the finished polymer can be represented as a structural formula with the specification of the degree of polymerization. This indicates the number of monomer units present in the polymer chain and is thus a measure of the molar mass of the polymer. The degree of polymerization can be controlled via the reaction conditions, the number of monomers, initiators or by the choice of a catalyst. The molar mass distribution of a finished polymer is in turn decisive for the resulting properties, as shown in Section 1.3.1.

Processing of Plastics

The plastics obtained either by polymerization as virgin material or by processing of waste plastics as recyclates are available for a wide variety of processing operations. In the recycling sector, so-called compounding systems are used for processing and, in general, for modifying plastics. The production of plastic products can then be carried out, for example, using the standard processes of extrusion or injection molding.

The compounding system

In a compounding system, plastics are provided with additives; therefore, the mixing performance of such a system is the decisive process criterion. The co-rotating, closely intermeshing twin-screw extruder is available as the standard machine for this task. Due to the very different mixing tasks, it has a modular design (see Figure 1.17). The processing unit consists of barrel elements that are connected to one another via tie rods. The two closely intermeshing screws each consist of a screw mandrel onto which the functional screw elements are fitted. Through the interaction of conveying, kneading, mixing and other special elements, the appropriate screw can be assembled for the incorporation of a wide variety of additives. The polymer is fed in via the main hopper, while the additives are usually metered in via side feeders along the screw. In most cases, the material is also degassed during processing to remove volatile components. At the end of the compounding section, the material is converted into the usual pellet form by a pelletizing unit.

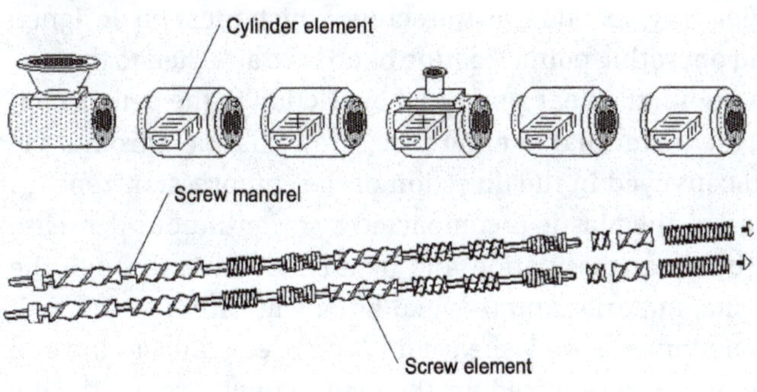

Figure 1.17 Modular design of a co-rotating twin-screw extruder [Sch21, p. 67]

The extrusion process

Extrusion lines are used for the production of semi-finished products with a constant cross-section in a continuous process. Using the example of a pipe extrusion line, the basic, exemplary design of such a line will be shown (see Figure 1.18). The respective plastic is fed as free-flowing pellets or as powder via a hopper to the extruder (1), conveyed in the direction of the screw tip and melted by shear dissipation and heat conduction. The resulting melt is forced through a pipe die (2), which gives the melt the shape of an annular ring. The strand emerging from the die enters the calibration unit (3), in which the plastic is frozen under dimensional constraint. This is followed by a cooling section that further cools the product. A caterpillar or belt haul-off (5) ensures that the pipe is drawn from the calibrator at a defined speed. Since the extrusion of pipes is a continuous process and thus produces an endless product, a suitable cutting unit must be provided in the form of a traveling cutting device, e.g. a saw (7). The production of profiles is carried out with a basically identical line configuration, although the die is often considerably more complex in design.

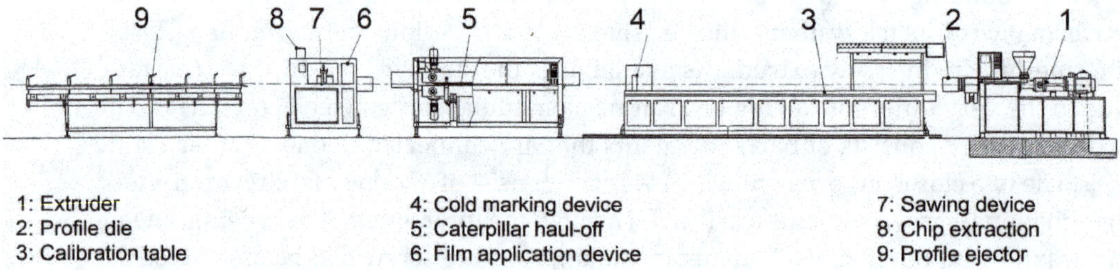

Figure 1.18 Pipe/profile extrusion line [Sch21, p. 19]

The first important step in the extrusion process is the provision of a materially homogeneous and temporally constant melt flow. For this purpose, single-screw extruders are used in most cases, the design of which is shown as an example in Figure 1.19. The central component of the single-screw extruder is the screw, which must be designed specifically for the material and operating point. Heating bands are attached to the surrounding barrel, which together with the heat generated by friction in the pellets melt the material. In the feed zone, the material that enters the screw channel through the inlet opening is picked up and conveyed in the direction of the compression zone. In the compression zone of the screw, the plastic is compacted by a continuous decrease in channel depth and melted by shear dissipation and heat conduction through the barrel wall. In the metering zone, material and thermal homogenization of the melt takes place. This is supported in many cases by shear and mixing elements at the end of the screw. Furthermore, the pressure required for the melt to flow through the die downstream of the extruder is built up in this zone.

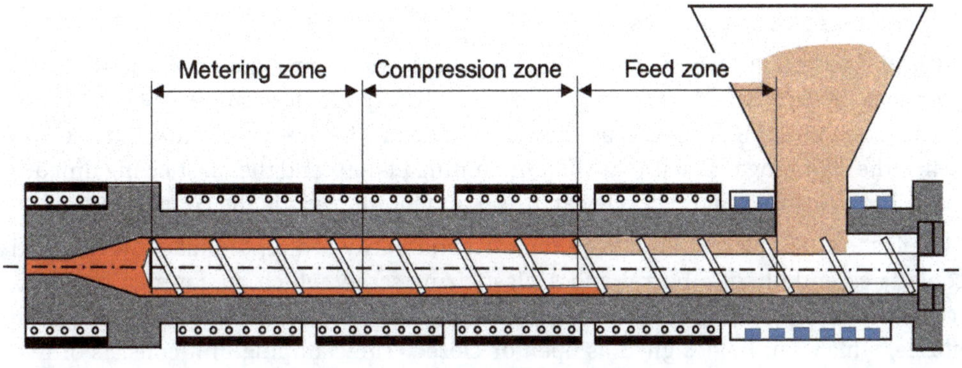

Figure 1.19 Functional zones of a three-zone screw [Sch22, p. 22]

The plastic melted in the extruder is reshaped in the pipe die (see Figure 1.20) from a mostly circular strand to an annular gap cross-section. This is achieved, for example, by a so-called spider die, in which the incoming circular melt stream flows around a mandrel (3) and is thus formed as an annular gap. In the area of the spider (2), which usually has the largest diameter, the melt is divided into several partial streams and flows around the spider legs (6), which fix the mandrel in the melt channel and hold it against the flow resistance. Behind the spider, the flow channel tapers and ends in the land zone (1) with parallel channel walls.

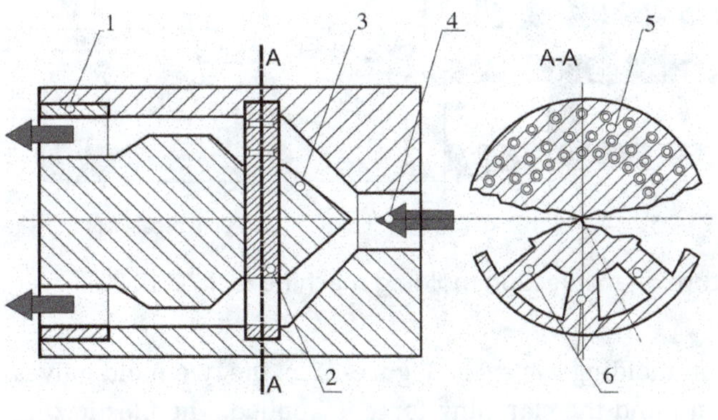

Figure 1.20 Longitudinal section of a spider die with spider leg or breaker plate as mandrel holder [GLF17, p. 79]

Instead of a pipe die, so-called flat film dies are used for the production of flat films, and the downstream equipment essentially consists of chill rolls, edge trimming and the winder. Films can also be produced using a blown film line. In this process, a tube is first extruded in the vertical direction and drawn upward. The film is stretched transversely by a slight overpressure. At the upper end of the film tower, the haul-off unit is located, which lays the tube flat and guides it back down to the winder.

The injection molding process

While the extrusion process is a continuous process for the production of semi-finished products, injection molding ejects one (or more) components at the end of each cycle. An injection molding machine, as shown in Figure 1.21, consists of a machine bed on which the clamping unit with the mold-mounting platens and the plasticizing unit are built. The injection molding cycle can be set and monitored via the control system.

The plasticizing unit can be moved forward and backward on the machine bed and consists essentially of a barrel and a plasticizing screw which, in contrast to an extrusion screw, additionally features a non-return valve. At the end of the barrel is the nozzle, which can be designed as open or closed. The clamping unit consists of two platens onto which the injection mold (negative form of the component) is clamped and which are located in a mechanically coupled system for applying the clamping force. The force with which the mold is held closed during injection can be applied either via a hydraulic piston or via a toggle lever system.

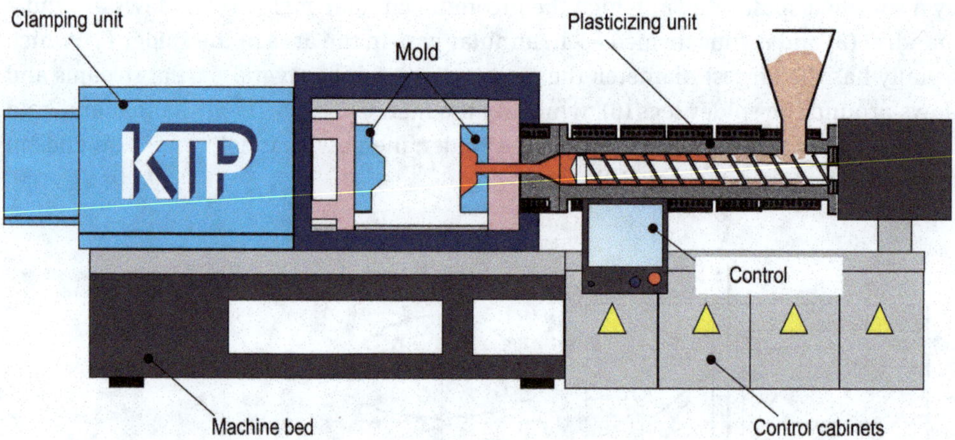

Figure 1.21 Schematic structure of an injection molding machine [Sch22, p. 32]

At the beginning of an injection molding cycle (see Figure 1.22), the two mold halves are closed via the clamping unit and the clamping force is applied. The plasticizing unit moves forward to the sprue bushing of the mold and injects the previously metered material into the injection mold. For this purpose, the injection molding screw performs a translational movement, whereby the non-return valve closes and backflow of the melt into the screw flights is prevented. After the volumetric filling of the mold cavity accomplished by the injection, the holding pressure, again applied by the screw, takes effect. This causes material to be packed into the cavity until the melt solidifies in the sprue bushing, in order to counteract the volumetric shrinkage caused by cooling. After the end of the holding pressure phase, the component continues to cool while the plasticizing unit moves back a short distance and material is metered again for the next cycle. For this purpose, the screw rotates, drawing in granulate and melting the plastic through dissipation and heat conduction. The melt is conveyed into the screw antechamber, whereby the screw is pushed backward while maintaining a set back pressure. At the end of the metering process, the quantity of plastic melt required for the next injection process is available in the screw antechamber. On the clamping side, after the end of the residual cooling time, the mold opens and the components are ejected.

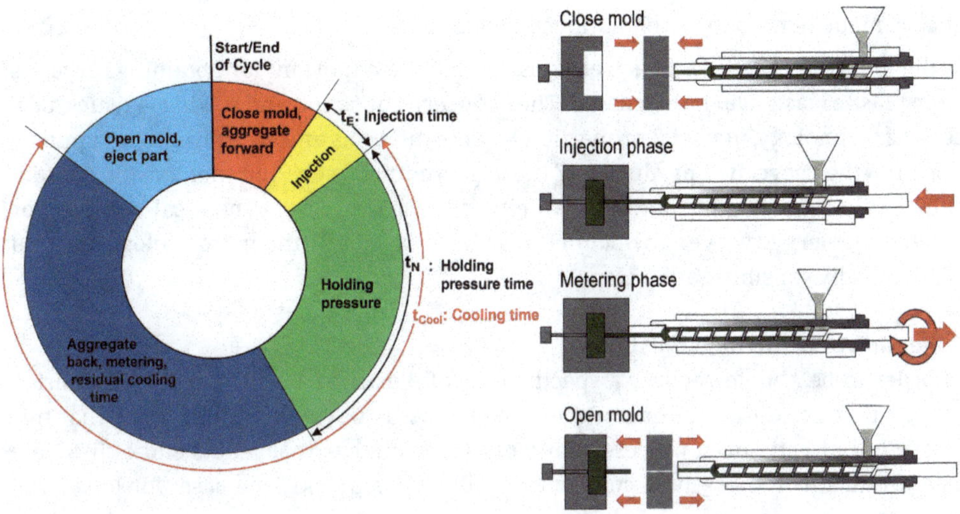

Figure 1.22 Sequence of the injection molding cycle [Sch22, p. 34 ff.]

For the production of complex components, numerous process variants are available, such as co-injection molding for the material-bonded joining of two or more plastic materials, sandwich injection molding for the production of a component with a skin and core layer, or back injection molding for back-injecting decorative films or textiles. Further variants integrate surface activation with plasma ("InMold-Plasma") [Nor18] or the injection of gas for the production of hollow structures into the injection molding process [Plu14; Sei16; Lan19].

1.3.3 From Waste to Recyclable Material

The requirements placed on the raw material for the production of plastic components vary depending on the application. These have been fully met for decades through the use of virgin plastic material and by modifying it. Virgin material is available as pure material in high delivery quantities and with high delivery reliability at sufficient quality. The need to use recyclate is therefore not present at all in many applications.

Many processors ask themselves why they should use a material that sometimes fails to meet important requirements and is frequently at a disadvantage in price competition against virgin material.

Nevertheless, recyclate has been used in numerous applications for years with an increasing trend. On the one hand, this is because more and more products are being identified for which the use of recyclate is feasible from a technological and economic perspective. On the other hand, the social and political pressure on manufacturers, processors and retailers to adopt a more sustainable approach is increasing. Reference is made here to the explanations in chapter 2.

In the following, fundamental prerequisites that are important for obtaining recyclate from plastic waste will be discussed. They concern collection and sorting logistics, processing technology, material properties, legal regulations and the existence of applications for the recyclates produced. If these prerequisites are not met, or are only insufficiently met, the plastic waste can either be processed via chemical recycling or it is fed to energy recovery, which in these cases is usually the more ecologically and economically sensible option.

Collection and Sorting Logistics

In order to be able to recover a specific type of plastic as a secondary raw material, it must first be present in one of the numerous waste streams. This means that this plastic has already been processed into products and has reached its end-of-life state as a product. For short-lived consumer goods this is quickly the case; for long-lived products, when new types of plastic are introduced, this can take a longer time. This applies, for example, to insulation materials in the construction sector, which, when they appear in the waste stream, may already be over 40 years old.

The next prerequisite is that the desired plastic reliably enters one of the organized collection systems. In the area of end consumers, this is, for example, the collection of lightweight packaging via the recyclables bin or the collection of PET bottles via the deposit system. Through B2B business relationships, specific plastic waste can also be sourced and processed directly.

Sorting technology is the next step for separating the desired plastic from the waste stream. Only if suitable sorting processes are available in the respective plants does the possibility arise of recovering specific plastics. The quantity of plastic in the waste stream plays a significant role with regard to the economical operation of the plant. Much is technically possible, but considerably less is economically feasible (see 3.1.7). The classic recyclate plastics such as PE or PP are present in large quantities in the lightweight packaging waste stream and are reliably separated. Specialty grades of polyamides, for example, end up in the mixed plastics fraction in a light packaging sorting plant and are mostly sent for energy recovery.

Processing Technology

If the desired plastic has been successfully assigned, consciously or unconsciously (as an interfering or foreign material), to at least one sorting fraction in the sorting plant, the next challenge lies in the separation of the waste pieces and the separation of the plastic type from the remaining waste stream. For this purpose, suitable size reduction, cleaning and separation technology must be available, e.g. at a recyclate manufacturer. The different plant components are described in chapter 3. Here again, the question of economic viability for the separation and removal of the desired plastic type arises. For the commodity plastics HDPE, LDPE or PP, large-scale plants with high throughput are state of the art.

Material Properties and Degradation

If the desired plastic can be separated from the waste stream, the question of its properties compared to those of the virgin material possibly used until then for a specific product still remains. This concerns both the processing properties and the properties in use. It must be taken into account that every product has a specific requirement profile for the material. For each type of plastic, there is therefore a large number of individual variants on the virgin material market that differ in numerous properties. There are homo- and copolymers, uncolored and colored, impact-modified, reinforced or filled and unfilled grades, each of which also exhibits differences in rheological behavior.

In the waste stream there is always a large mixture of many different variants of the same plastic type, whose separability is severely limited. Thus, filled and unfilled variants can under certain circumstances be separated, but the adjustment of rheological properties takes place in the course of the specific processing of the recyclates, if this is economically feasible.

In addition to the variety of variants of individual types of a plastic in the waste stream, the aspect of degradation of the plastics plays a major role in assessing the usability of a recyclate in place of virgin material. Plastics are subject to numerous potentially damaging influences during processing and use. Irreversible damage to plastics manifests itself at the molecular level in a change in molecular structure (chain scission, crosslinking, branching), in the formation of functional groups and in the cleavage of chain segments [EP07, p. 30].

Typical influencing factors for such damage are:

- Related to the processing and reprocessing of plastics:
  - The processing temperature (thermal degradation)
  - The shear and elongational stresses introduced by rotating machine parts (thermo-mechanical degradation)
  - The presence of atmospheric oxygen (thermo-oxidative degradation)
  - The presence of moisture in the plastic melt (hydrolysis)

- Related to the use phase:
  - UV radiation
  - Temperature
  - Humidity, water (hydrolysis)
  - Oxygen
  - Seawater (salt, mechanical action through wave motion)
  - Chemicals
  - High-energy radiation
  - Gaseous pollutants
  - Biological influencing factors (microorganisms)

How strongly the aforementioned influencing factors affect the actual degree of damage depends in turn on the chemical and physical structure of the plastics, on the fillers present in the plastic and on possible polymerization residues still present in the plastic [EP07, p. 30–84].

Chain scission through degradationprocesses cause in particular an increase in the melt flow index MFI (Melt Flow Index), which is a measure of the flowability of the plastic melt. The higher this value, the lower the viscosity of the material. In addition, the impact strength of the plastic decreases, while the tensile strength only decreases with very severe material degradation. Figure 1.23 shows the stress-strain behavior for a polypropylene and the progression of the MFI value after five passes through the processing steps of part production by injection molding, size reduction and regranulation. The tensile strength remains largely constant, while the MFI value increases significantly with the number of cycles. This shows that repeated processing alone causes significant thermo-mechanical degradation, without even taking into account the use phase with its additional influencing factors.

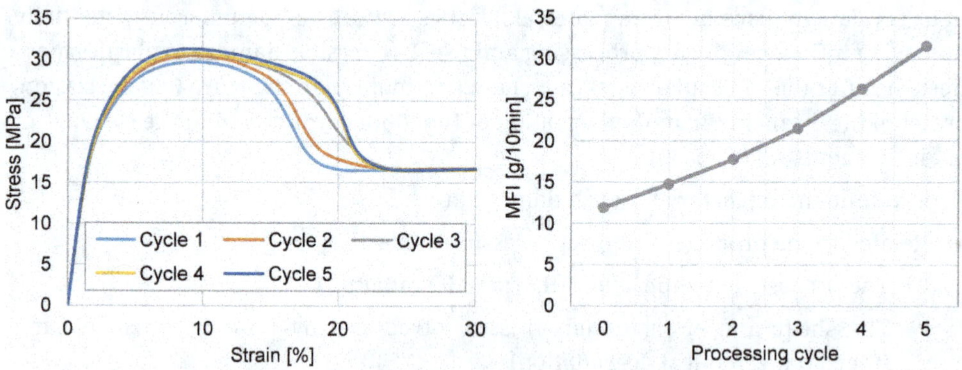

Figure 1.23 Progression of stress-strain behavior (left) and MFI value (230 °C/ 2.16 kg) (right) over five processing cycles for a polypropylene homopolymer

In order to account for the strongly batch-dependent property differences of the cleaned regrind, virgin material regrind is frequently blended in. For this purpose, however, the regrind should be available as a single-grade material. Specially processed recyclate is in many cases not worthwhile for economic reasons, although PET, for example, is an exception. Here, a complex processing plant is state of the art. Single-grade and clean plastic waste, such as that occurring in inline recycling in production plants or in post-industrial waste, can frequently be processed without major losses in the property profile.

Mixed plastic waste is in most cases sent for energy recovery, because the different plastics are in the rarest of cases directly compatible with one another. Although additives can be used that can increase compatibility, the analysis of the waste stream and the selection of suitable additives are complex. Products from the processing of these regrind materials exhibit significantly inferior mechanical properties compared to virgin material.

Applications

To return to the example of a plastic to be recovered from the waste stream, the final step of manufacturing a new product follows after the assessment of the material properties. For single-grade post-industrial recyclates (PIR) and for regrind from inline recycling, there are hardly any restrictions with regard to product design. Processed regrind or recyclate from the post-consumer waste sector, however, require a more detailed examination of the properties to be achieved in conjunction with the adaptation of the product design ("design for recyclates") and the modification of the processing equipment. For the processing of mixed plastic waste, several applications have been developed over the past decades through robust product design, such as park benches, step boards, or fence elements.

For the most important plastics in terms of recycling, Table 1.3 lists some typical areas of application for recyclates.

Table 1.3 Overview of areas of application for recyclates [Poh20, p. 470 f.; NN03]

Polyethylene (PE)	Agricultural films, packaging films, fence elements, crates, bottles, furniture accessories, containers, barrier grids/bollards, stationery, flower pots, cable protection, drainage systems, road speed bumps
Polypropylene (PP)	Automotive components, pallets, containers, housings, furniture accessories, folding boxes, laundry baskets, household appliances, drainage systems, soakaway boxes
Polyethylene terephthalate (PET)	PET bottles, textile fibers, films, boat sails, automotive parts
Polyvinyl chloride (PVC)	Window profiles, pipes, floor coverings, traffic cones

Legal Regulations

In order to be able to use plastic waste again as a secondary raw material, it is necessary, in addition to technical and economic feasibility, to examine whether regulations must be observed depending on the composition of the recyclate or the application.

For recyclates that come into direct contact with food, there are strict regulations governed by the European Food Safety Authority (EFSA) of the EU. All substances contained in the recyclate must be registered and the recycling process must receive approval from the EFSA (see 4.4.2).

In principle, the provisions of the REACH regulation apply to recyclates, as they do to virgin plastics (see 4.4.1). In other industries, additional strict testing requirements must also be met, such as in the electronics or toy industry. In addition to pollutants that were still used as additives some time ago and could today find their way into recyclates, contamination or foreign substances are frequently a knock-out criterion for use in many areas of application.

On the other hand, legal requirements regarding a recyclate quota, e.g. in PET bottles, help to ensure that the processing of the recyclate can be implemented economically. For the required proportion, there is a demand for which virgin material is not in competition. This promotes the production and use of recyclate from plastic waste.

Summary

The production of recyclate that can compete with virgin material from a technical and economic point of view requires, in most cases, a single-grade material stream with no or only minor contamination. Otherwise, sorting and washing would be too cost-intensive and the recyclate produced could only be offered at a higher price compared to virgin material.

For applications with low requirements, recyclates from material streams with a high degree of contamination and low grade purity have found a niche that has been served for many years. The material input is often higher due to the lower mechanical properties, i.e., the products manufactured are built more robustly.

Material recycling is therefore not possible for all types of waste. The respective quantities of waste play a major role here. The high proportion of packaging waste from the food sector, which is to be counted among short-lived consumer goods, cannot, for example, be reused as food packaging due to legal regulations, with few exceptions. This issue is discussed in more detail in 2.4.

Since feedstock recycling frequently places lower demands on the purity of the waste stream, it represents an alternative with regard to material recovery for basic chemicals in the chemical industry. Due to the high calorific value, plastic waste that cannot be recycled as material is also of interest for energy recovery. For further details regarding the three recovery pathways of material recycling, feedstock recycling, and energy recovery, reference is made to chapters 4 to 6.

References

[AAP+19] Allen, S.; Allen, D.; Phoenix, V.; Le Roux, G.; Jimenez, P.; Simmonneau, A.; Binet, S.; Galop, D.: "Atmospheric transport and deposition of microplastics in a remote mountain catchment". nature geoscience, 2019

[BBB+21] Bertling, J.; Bannick, C. G.; Brinkmann, L.; Barkmann, L.; Braun U.; Knoblauch, D.; Kraas, C.; Mederake, L.; Nosić, F.; Philipp, B.; Sartorius, I.; Schritt, H.; Stein, U.; Wencki, K.; Wendt-Potthoff, K; Woidasky. J.: „Kunststoff in der Umwelt – ein Kompendium" Hrsg.: Ecologic Institut gemeinnützige GmbH, 1. Auflage, 2021

[BBH+18] Bertling, J.; Bertling, R.; Hamann, L.: „Kunststoffe in der Umwelt: Mikro- und Makroplastik. Ursachen, Mengen, Umweltschicksale, Wirkungen, Lösungsansätze, Empfehlungen". Kurzfassung der Konsortialstudie, Fraunhofer-Institut für Umwelt-, Sicherheits- und Energietechnik UMSICHT (Hrsg.), Oberhausen, Juni 2018

[BBO+13] Baur, E.; Brinkmann, S.; Osswald, T.; Rudolph, N.; Schmachtenberg, E.: „Saechtling Kunststoff Taschenbuch". Carl Hanser Verlag, München, 2013

[BDS+22] Bertling, J.; Dau, K.; Selig, U.; Werner, S.: "Release of Microplastics into the Marine Environment – State of Knowledge and Options for Action". German Roundtable on Marine Litter, DOI: https://doi.org/10.24406/umsicht-n-648144

[BF17] Boucher, J.; Friot, D.: "Primary Microplastics in the Oceans – a Global Evaluation of Sources". IUCN, Gland, Switzerland, 2017

[Bra13] Braun, D.: „Kleine Geschichte der Kunststoffe". Carl Hanser Verlag, München, 2013

[Con21] Conversio Market & Strategy GmbH: „Stoffstrombild Kunststoffe in Deutschland 2021", November 2022

[CEG+14] Cózar, A.; Echevarria, F.; González-Cordillo, J. I.; Irigoien, X.; Ubeda, B.; Hernandez-Leon, S.: "Plastic debris in the open ocean". In: Proceedings of the National Academy of Sciences of the United States of America, Jhrg. 111, Vol. 28, pp. 10239–10244, DOI: 10.1073/pnas.1314705111

[DGM+17] Dris, R.; Gasperi, J.; Mirande, C.; Mandin, C.; Guerrouache, M.; Langlois, V.; Tassin, B.: "A first overview of textile fibers, including microplastics, in indoor and outdoor environments". In: Environmental Pollution, Vol. 221, pp. 453–458, 2017

[DGS+16] Dris, R.; Gasperi, J.; Saad, M.; Mriande-Bret, C.; Tassin, B.: "Synthetic fibers in atmospheric fallout: A source of microplastics in the environment?". In: Marine Pollution Bulletin, Vol. 104, issues 1–2, pp. 290–293, 2016

[DHM22] Dahlmann, R.; Haberstroh, E.; Menges, G.: „Menges Werkstoffkunde Kunststoffe". Carl Hanser Verlag, München, 2022

[ELC+14] Eriksen, M.; Lebreton, L. C. M.; Carson, H. S.; Thiel, M.; Moore, C. J.; Borerro, J. C.; Galgani, F.; Ryan, P. G.; Reisser, J.: "Plastic Pollution in the World's Oceans: More than 5 Trillion Plastic Pieces Weighing over 250,000 Tons Afloat at Sea". In: PLoS ONE, Vol. 12 (9), DOI: 10.1371/journal.pone.0111913, 2014

[EP07] Ehrenstein, G. W.; Pongratz, S.: „Beständigkeit von Kunststoffen", Band 1, Carl Hanser Verlag, München, 2007

[GCH+21] González-Fernández, D.; Cózar, A.; Hanke, G.; Viejo, J.; Morales-Caselles, C.; Bakiu, R.; Barceló, D.; Bessa, F.; Bruge, A.; Cabrera, M.; Castro-Jiménez, J.; Constant, M.; Crosti, R.; Galletti, Y.; Kideys, A. E.; Machitadze, N.; Pereira de Brito, J.; Pogojeva, M.; Ratola, N.; Rigueira, J.; Rojo-Nieto, E.; Savenko, O.; Schöneich-Argent, R. I.; Siedlewicz, G.; Suaria, G.; Tourgeli, M.: "Floating macrolitter leaked from Europe into the ocean". In: Nature Sustainability, Vol. 4 (6), pp. 474–483, DOI: 10.1038/s41893-021-00722-6, 2021

[Gey20] Geyer, R.: "Production, use, and fate of synthetic polymers". In: Plastic Waste and Recycling, Edited by Trevor M. Letcher, Elsevier, London, 2020

[GJL17] Geyer, R.; Jambeck, J.; Law, K. L.: "Production, use, and fate of all plastics ever made". In: Science Advances, Vol. 3, DOI: 10.1126/sciadv.1700782, 2017

[GKR20] Gabbott, S.; Key, S.; Russell, C.; Yonan, Y.; Zalasiewicz, J.: "The geography und geology of plastics: their environmental distribution and fate". In: Plastic Waste and Recycling, Edited by Trevor M. Letcher, Elsevier, London, 2020

[GLF17] Greif, H.; Limper, A.; Fattmann, G.: „Technologie der Extrusion". Carl Hanser Verlag, München, 2017

[HB12] Hoornweg, D.; Bhada-Tata, P.: "What a waste: A global Review of Solid Waste Management". Urban Development Series, World Bank, 2012

[HSW+17] Horton, A. A.; Svendsen, C.; Williams, R. J.; Spurgeon, D. J.; Lahive, E.: "Large microplastic particle in sediments of tributaries of the River Thames, UK – abundance, sources and methods for effective quantification". In: Marine Pollution Bulletin, Vol. 114, issue 1, pp. 218–226, 2017

[JGW+15] Jambeck, J. R; Geyer, R.; Wilcox, C.; Siegler, T. R.; Perryman, M.; Andrady, A.; Narayan, R.; Law, K. L.: "Plastic waste inputs from land into the ocean". In: Science, Vol. 347, issue 6223, 2015

[Kai16] Kaiser, W.: „Kunststoffchemie für Ingenieure – Von der Synthese bis zur Anwendung". Carl Hanser Verlag, München, 2016

[Ker16] Kershaw, P. J. et al.: "Marine plastic debris and microplastics – Global lessons and research to inspire action and guide policy change". United Nations Environment Programme (UNEP), 2016

[Krü12] Krüger, O.: „Kunststoffe". In: „Werkstoffkunde". Bargel, H.-J.; Schulze, G. (Hrsg.), Springer-Verlag, Berlin, Heidelberg, 2012

[LA19] Lebreton, L.; Andrady, A.: "Future scenarios of global plastic waste generation and disposal". In: Palgrave Communications, Vol. 5, Article 6, 2019

[Lan19] Landgräber, B.: „Experimentelle und modellbasierte Analysen der Prozessphasen des Spritzgießsonderverfahrens GITBlow". Dissertation, Universität Paderborn, Kunststofftechnik Paderborn, 2019

[LMG+14] Law, K. L.; Moret-Ferguson, S.; Goodwin, D. S.; Zettler, E. R.; De-Force, E.; Kukulka, T.; Proskurowski, G.: "Distribution of surface plastic debris in the eastern Pacific Ocean from an 11-year dataset". In: Environ. Sci. Technol. 48, pp. 4732–4738, 2014

[LLS+18] Liu, M.; Lu, S.; Song, Y.; Lei, L.; Hu, J.; Lv, W.; Zhou, W.; Cao, C.; Shi, H.; Yang, X.; He, D.: "Micro-plastic and mesoplastic pollution in farmland soils in suburbs of Shanghai, China". In: Environmental Pollution Vol. 242, pp. 855–862, 2018

[LSF+18] Lebreton, L.; Slat, B.; Ferrari, F.; Sainte-Rose, B.; Aitken, J.; Marthouse, R.; Hajbane, S.; Cunsolo, S.; Schwarz, A.; Levivier, A.; Noble, K.; Debeljak, P.; Maral, H.; Schoeneich-Argent, R.; Brambini, R.; Reisser, J.: "Evidence that the Great Pacific Garbage Patch is rapidly accumulating plastic". In: Nature/ Scientific Reports, Vol. 8, issue: 4666, DOI:10.1038/s41598-018-22939-w, 2018

[LSL+17] Ling, S.; Sinclair, M.; Levi, C.; Reeves, S.; Edgar, G.: "Ubiquity of microplastics in coastal seafloor sediments". In: Marine Pollution Bulletin Vol. 121, pp.104–110, 2017

[LZD+17] Lebreton, L. C. M.; van der Zwet, J.; Damsteeg, J.-W.; Slat, B.; Andrady, A.; Reisser, J.: "River plastic emissions to the world's oceans". In: Nature Communications, Vol. 8, issue:15611, DOI: 10.1038/ncomms15611, 2017

[Mck16-ol] McKinsey & Company, Inc.: Globale Kunststoff- und Plastikmüllproduktion 2016, https://www.mckinsey.de/~/media/mckinsey/locations/europe%20and%20middle%20east/deutschland/news/presse/2018/2018-12-19-plastikmuell/kunststoff-%20und%20plastikmuellproduktion%20global_heute%20und%202030.pdf, zuletzt abgerufen am 15.10.2021

[MF11] Milliman, J. D.; Farnsworth, K. L.: "River Discharge to the Coastal Ocean: A Global Synthesis". Cambridge University Press: Cambridge, 2011.

[MKZ+17] Machado, A. A. d. S.; Kloas, W.; Zarfl, C.; Hempel, S.; Rillig, M. C.: "Microplastics as an emerging threat to terrestrial ecosystems". In: Global Change Biology, Vol. 24, pp. 1405–1416, 2017

[MVG+21] Morales-Caselles, C.; Viejo, J.; Marti, E.; González-Fernández, D.; Pragnell-Raasch, H.; González-Gordillo, J. I.; Montero, E.; Hanke, G.; Salvo, V. S.; Basurko, O. C.; Mallos, N.; Lebreton, L.; Echevarría, F.; van Emmerik, T.; Duarte, C. M.; Gálvez, J. A.; van Sebille, E.; Galgani, F.; García, C. M.; Ross, P. S.; Bartual, A.; Ioakeimidis, C.; Markalain, G.; Isobe, A.; Cózar, A.: "An inshore–offshore sorting system revealed from global classification of ocean litter". In: Nature Sustainability. Vol. 4 (6), pp. 484–493, DOI: 10.1038/s41893-021-00720-8, 2021

[NN01-ol] Ocean Conservancy, McKinsey Center for Business and Environment: Stemming the tide: land-based strategies for a plastic-free ocean, 2015, https://www.mckinsey.com/~/media/mckinsey/business%20functions/sustainability/our%20insights/saving%20the%20ocean%20from%20plastic%20waste/stemming%20the%20tide%20full%20report.ashx, abgerufen am: 09.08.2022

[NN02-ol] "Plastics – the facts 2022 – An analysis of European plastics production, demand and waste data". PlasticsEurope, 2022, https://plasticseurope.org/de/wp-content/uploads/sites/3/2022/10/PE-PLASTICS-THE-FACTS_20221017.pdf, abgerufen am: 23.08.2023

[NN03-ol] Westdeutscher Rundfunk Köln: „Destillation von Erdöl". https://www.planet-wissen.de/technik/energie/erdoel/index.html, abgerufen am 01.07.2022

[Nor18] Nordmeyer, T.: „Verfahrenstechnische Entwicklung des Direktinjektion-Plasmaverfahrens im Spritzgießprozess". Dissertation, Universität Paderborn, Kunststofftechnik Paderborn, 2018

[NT16] Napper, I. E.; Thompson, R. C.: "Release of synthetic microplastic plastic fibres from domestic washing machines: Effects of fabric type and washing conditions". In Marine Pollution Bulletin, Vol. 112, Issue 1–2, pp. 39–45, 2016

[Plu14] Plugge, T: „Herstellung komplexer polymerer Hohlkörperstrukturen – Ganzheitliche Prozessanalyse und -optimierung des Spritzgießsonderverfahrens GITBlow". Universität Paderborn, Kunststofftechnik Paderborn, 2014

[Poh20] Pohjakallio, M.: "Secondary plastic products – examples and market trends". In: Plastic Waste and Recycling, Edited by Trevor M. Letcher, Elsevier, London, 2020

[PPB+19] Peeken, I.; Primpke, S.; Beyer, B.; Gütermann, J.; Katlein, C.; Krumpen, T.; Bergmann, M.; Hehemann, L.; Gerdts, G.: "Arctic sea ice is an important temporal sink and means of transport for microplastic". Nature Communications Vol. 9, issue:1505, DOI: 10.1038/s41467-018-03825-5, 2019

[Rei21] Reich, O.: "The changing additivation needs in a circular economy". Presentation on 3rd Forum plastics recyclate, 09.03.2021

[RH18] Rist, S.; Hartmann, N. B.: "Aquatic Ecotoxicity of Microplastics and Nanoplastics: Lessons Learned from Engineered Nanomaterials". In: Freshwater Microplastics, Editors: Wagner, M.; Lambert, S., Springer Nature, Vol. 58, DOI: 10.1007/978-3-319-61615-5_2, 2018

[RSN+15] Reisser, J.; Slat, B.; Noble, K.; Du Plessis, K.; Epp, M.; Proietti, M.; Sonneville, J. de; Becker, T.; Pattiaratchi, C.: "The vertical distribution of buoyant plastics at sea: an observational study in the North Atlantic Gyre". In: Biogeosciences, Vol. 4 (12), pp. 1249–1256, DOI: 10.5194/bg-12-1249-2015, 2015

[Sch21] Schöppner, V.: „Standardverfahren Extrusion". Vorlesungsskript, Universität Paderborn, Kunststofftechnik Paderborn, 2021

[Sch22] Schöppner, V.: „Grundlagen der Kunststoffverarbeitung". Vorlesungsskript, Universität Paderborn, Kunststofftechnik Paderborn, Sommersemester, 2022

[Sei16] Seidel, S.: „Prozessphasenoptimierte Simulation und Modellierung des Spritzgießsonderverfahrens GITBlow". Dissertation, Universität Paderborn, Kunststofftechnik Paderborn, 2016

[She16-ol] Sherrington, C.: "Plastics in the Marine Environment". Eunomia, Bristol, United Kingdom, 2016, https://safety4sea.com/wp-content/uploads/2016/06/Eunomia-Plastics-in-the-Marine-Environment-2016_06.pdf, abgerufen am: 09.08.2022

[SRH+20] Schernewski, G.; Radtke, H.; Hauk, R.; Baresel, C.; Olshammar, M.; Osinski, R.; Oberbeckmann, S.: "Transport and Behavior of Microplastics Emissions from Urban Sources in the Baltic Sea". In: Frontiers in Environmental Science, Vol. 8, 2020

[SRH+21] Schernewski, G.; Radtke, H.; Hauk, R.; Baresel, C.; Olshammar, M.; Oberbeckmann, S.: "Urban Microplastics Emissions: Effectiveness of Retention Measures and Consequences for the Baltic Sea". In: Frontiers in Marine Science, Vol. 8, 2021

[SKW17] Schmidt, C.; Krauth, T.; Wagner, S.: "Export of Plastic Debris by Rivers into the Sea". In: Environmental science & technology, Vol. 51 (21), pp. 12246–12253, DOI: 10.1021/acs.est.7b02368, 2017

[SWL+15] van Sebille, E.; Wilcox, C.; Lebreton, L.; Maximenko, N.; Hardesty, B. D.; van Franeker, J. A.; Eriksen, M.; Siegel, D.; Galgani, F.; Law, K.: "A global inventory of small floating plastic debris". In: Environmental Research Letters, Vol. 12 (10), DOI: 10.1088/1748–9326/10/12/124006, 2015

[SWL+18] Scherer, C.; Weber, A.; Lambert, S.; Wagner, M.: "Interactions of microplastics with freshwater biota". In: Freshwater Microplastics, Editors: Wagner, M.; Lambert, S., Springer Nature, Vol. 58, DOI: https://doi.org/10.1007/978-3-319-61615-5_8, 2018

[THR+19] Turner, S.; Horton, A. A.; Rose, N.; Hall, C.: "A temporal sediment record of microplastics in an urban lake, London, UK". In: Journal of Paleolimnology, Vol. 61, pp. 449–462, 2019

[VTR17] Vaughan, R.; Turner, S.; Rose, N.: "Microplastics in the sediments of a UK urban lake". In Environmental Pollution, Vol. 229, pp.10–18, 2017

[WL20] Welden, N. A.; Lusher, A.: "Microplastics: from origin to impacts". In: Plastic Waste and Recycling, Edited by Trevor M. Letcher, Elsevier, London, 2020

2 From Linear to Circular Management

As shown in Chapter 1, the rapid development of plastics over recent decades has led to this material being omnipresent for every person. On the positive side, this means, for example, pursuing lightweight construction through highly innovative products, saving transport weight through weight reduction and thus also CO_2 savings and much more. On the negative side, plastics are unfortunately also present in the environment, which is mainly attributable to the way humans handle plastic waste.

This has led to the socio-political discourse on this topic resulting in a multitude of opinions, political positions and derived measures. The core and starting point of the discussions in this chapter is a politically desired circular economy for plastics. First, the political background and the legal provisions derived from it are presented. Then the key stakeholders along the value chain of plastic products are examined with their interests, implementation options, dependencies and their significance. Finally, selected material flows with production and waste quantities and their respective characteristics are shown. It becomes clear here that circular management is easier to implement in some industries than in others and that there are currently still many obstacles to overcome.

2.1 Linear Management Vs. Circular Economy

The linear economy describes the handling of materials characterized by the fact that the material is disposed of as a no longer usable resource after the use phase. With regard to the plastics economy, this is represented as shown in Figure 2.1.

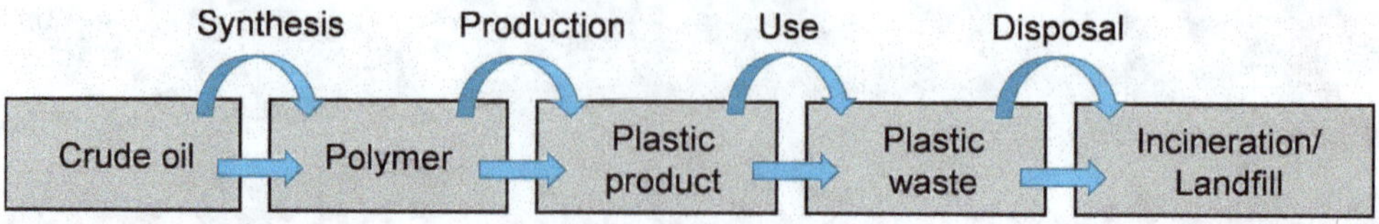

Figure 2.1 Linear economy of the plastics industry

From the raw material crude oil, the polymer is produced by distillation, steam cracking and a synthesis reaction. Through processing operations involving primary shaping, forming or joining processes, plastic components are produced, which are transferred to the use or service phase through sale. After the use phase, the components become plastic waste, which is disposed of or eliminated by incineration or landfilling. The principle of linear management is also called Take – Make – Waste" or "Cradle – to – Grave".

The challenge of the linear economy lies in ensuring the worldwide availability of the raw material crude oil and in the requirement for a globally established waste recovery system that also includes local, household-level waste collection.

The global economic development of recent decades has also had consequences for the plastics economy. The world population and overall prosperity have grown. The rapid development of plastics in all areas of life has led to a sharp increase in the production volume of plastics. At the same time, the total amount of waste and also the amount of plastic waste has increased. Waste disposal is critical in some countries due to a lack of collection and recovery structures. The cross-border shipping of waste has exacerbated the situation with regard to improper landfilling or open burning on land. From an economic perspective, there are moreover numerous dependencies regarding the availability of the raw material crude oil.

Overall, it can be said that the linear economy has long been criticized because valuable resources are successively eliminated and new raw materials are repeatedly used that are fundamentally not available in unlimited quantities. Environmentally compatible management as well as material and energy efficiency are buzzwords that led at the political level in the 1990s to approaches toward a circular economy. This was continuously developed further both at the national and EU level and has led to today's perspective, which focuses on the circular economy.

In a circular economy, the materials used (raw material or polymer) remain in a material cycle beyond the end-of-life state of the products manufactured from them. Losses from this material cycle (waste, process losses, emissions) and the extraction of primary raw materials from the environment (virgin plastics) should be avoided as far as possible.

Figure 2.2 shows the guiding principle of the circular economy as represented by the Federal Ministry for the Environment. In addition to numerous other subject areas, such as protection against environmental toxins and radiation and the diversity of animals and plants, the Federal Ministry for the Environment, Climate Action, Nature Conservation and Nuclear Safety also deals with the resource-conserving use of raw materials. Since most recycling processes involve losses, the use of primary resources remains unavoidable for the foreseeable future. However, it is important to establish ecologically and economically viable recycling processes that lead to the most reliable possible production of secondary raw materials.

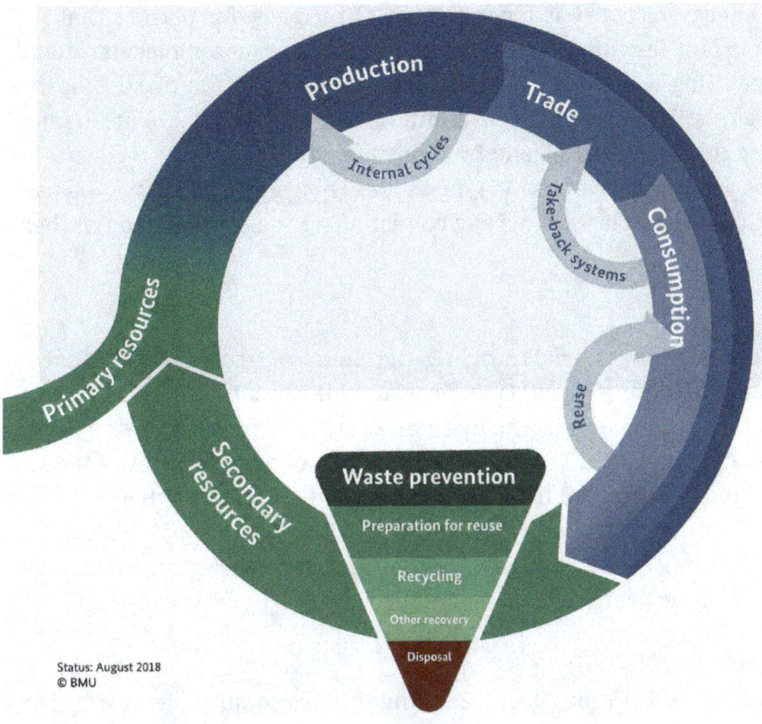

Figure 2.2 Guiding principle of the circular economy, [NN01-ol]

In addition, a contribution to conserving resources can be made by slowing down material cycles. This is achieved through a longer product service life. Reuse and preparation for reuse ("Repair", "Remanufacturing") are further building blocks that, in the case of plastics, can keep the material in a small cycle before it enters the large cycle of material recovery ("Recycle"). Internal cycles during production (inline recycling) or take-back systems (returnable packaging) are means of avoiding waste.

As shown in Figure 2.2, at the end of the consumption phase there is the so-called waste hierarchy of the EU Waste Framework Directive, which has been incorporated into the German Circular Economy Act. Accordingly, the appropriate waste treatment must be carried out for the waste generated in accordance with the order of priority. In doing so, it must be weighed up which of the options should be chosen taking into account the overall economic effort. In principle, the ecological and social costs, the costs of raw material extraction, the intergenerational and the business costs all play a role here. Against this background, in specific cases a loss of material from the cycle can contribute to the goals of the circular economy, for example when products containing pollutants are safely landfilled or incinerated rather than being kept in the cycle.

Energy Recovery

Incineration with energy recovery is a component of recovery for plastics that will continue to be important for the foreseeable future. Due to macromolecular degradation, material recycling leads to a long-term loss of quality of the plastic. Plastics therefore cannot be recycled indefinitely. In addition, composite components made of plastics and heavily contaminated plastic fractions are sent for energy recovery for economic as well as ecological reasons. As long as fossil energy carriers must be used for energy generation, waste plastics thereby contribute to the conservation of these fossil resources.

Material Recycling

When plastic waste is sent for material recycling, the chemical structure of the macromolecules is fundamentally preserved. Only unwanted degradation processes occur during the processing of the material. The main process steps are sorting, size reduction, cleaning, separation by plastic type and regranulation or agglomeration. Material recycling is also frequently referred to as physical recycling. However, there are also solvent-based processes that isolate the macromolecules but leave them intact. For this reason, in this work the mechanical and solvent-based processes are grouped together under the term material recycling.

Feedstock Recycling

Through chemical processes, in feedstock recycling the macromolecules are broken down to monomer level by chain scission. Monomers, oils or gases are produced, which can be used as a base material for polymer synthesis or for other chemical processes. Feedstock recycling is also frequently called chemical recycling and is used for mixed or contaminated plastic waste.

The circular economy is a politically desired economic system that will find its way into more and more industries. The production of plastic components will in future be based on the following raw material sources:

- Primary raw materials: virgin plastics
- Secondary raw materials: recyclate
- Processed semi-finished products

Consumers will use the following product categories:

- New products made from virgin material
- New products made from recyclate

- Used, reconditioned products
- Used, still intact products (secondhand)

In Section 2.2, the legal framework conditions that policy aims to implement for the transformation to a circular economy are presented.

2.2 Political Background and Legal Provisions

A decisive driver for the development toward a circular economy is the state in its legislative function. In this context, consumer behavior, the placing of products on the market, the use of raw materials and waste management are influenced and regulated. The role of the state as a stakeholder along the value chain within the framework of the circular economy is discussed in more detail in Section 2.3 and its influence on the other stakeholders is examined. In Section 2.2 the political background in the European Union with regard to waste and circular economy is first presented. This is followed by a description of some statutory provisions transposed from EU law into national law in Germany. Figure 2.3 provides an overview of selected directives and laws.

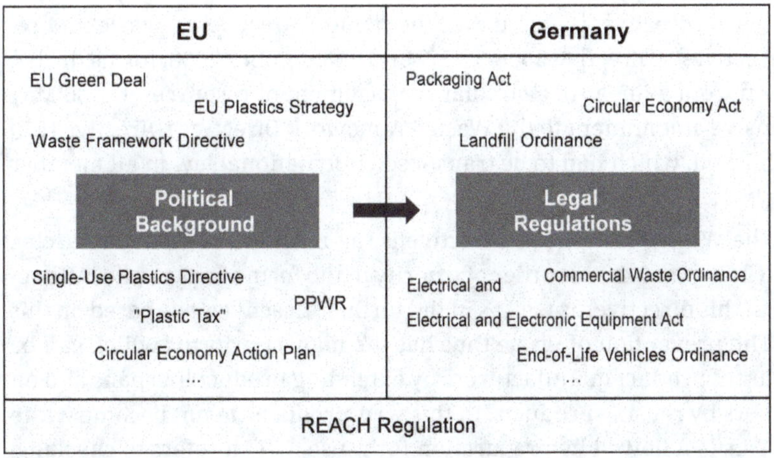

Figure 2.3 Selected EU directives as political background to the laws enacted in Germany on the subject of waste and circular economy

2.2.1 Political Background

From the Waste Framework Directive to the EU Green Deal with its resulting action plans, the European Union has been dealing with waste and circular economy and, in general, with environmental and climate issues for many years. The objectives of the directives can be summarized as follows:

- Protection of the environment and climate
- Protection of human health
- Improvement of resource use efficiency
- Promotion of the circular economy
- Reduction of waste generation

With the Waste Framework Directive, the first important regulations were established in 2008, forming the basis for the directives adopted from 2018 onwards. Beginning with the EU Plastics Strategy 2018, the EU has introduced a large number of regulations in recent years, demonstrating that the topic has gained significantly in importance. It is to be expected that in the coming years, member states will enact further regulations, prohibitions, quotas for recyclates, etc., by transposing EU requirements into national law in the form of legislation. The EU Packaging Regulation in 2025 is a further example of the tightening of legislative influence on industry and consumers with regard to waste and circular economy.

2.2.1.1 Waste Framework Directive

The Waste Framework Directive, issued under the name "Directive on waste and repealing certain Directives" (Directive 2008/98/EC) on 19 November 2008, forms the basis for the transition from a linear to a circular management of resources. On 30 May 2018, a comprehensive amendment to the Waste Framework Directive (Directive (EU) 2018/851) was published, which had to be transposed into national law in all member states within two years.

A key element of the Waste Framework Directive is the introduced waste hierarchy, which is presented in Figure 2.4 in its order of priority in the form of a pyramid. All provisions laid down in this directive, as well as in the further directives, are based on this order of priority. The prevention of waste thus has the highest priority, which can be achieved on the part of product manufacturers by extending product lifespans and on the part of consumers by reusing products. In this way, products do not become waste in the first place. This is followed by preparation for reuse, which refers to checking, cleaning or repairing products before their second life cycle, whereby waste status is reached prior to reuse.

Recycling includes material recycling and feedstock recycling, but not the processing of waste into substitute fuels. As part of the amendment to the directive in 2018, the term "material recovery" was additionally introduced, which encompasses preparation for reuse, recycling and also backfilling. Backfilling refers to the use of non-hazardous waste, e.g. in landscaping, whereby such waste must replace materials that are not waste. Other recovery refers to all further recovery operations that serve a useful purpose", meaning that they replace other materials otherwise intended to fulfil a specific function. Disposal refers to all further operations that serve waste treatment but not recovery.

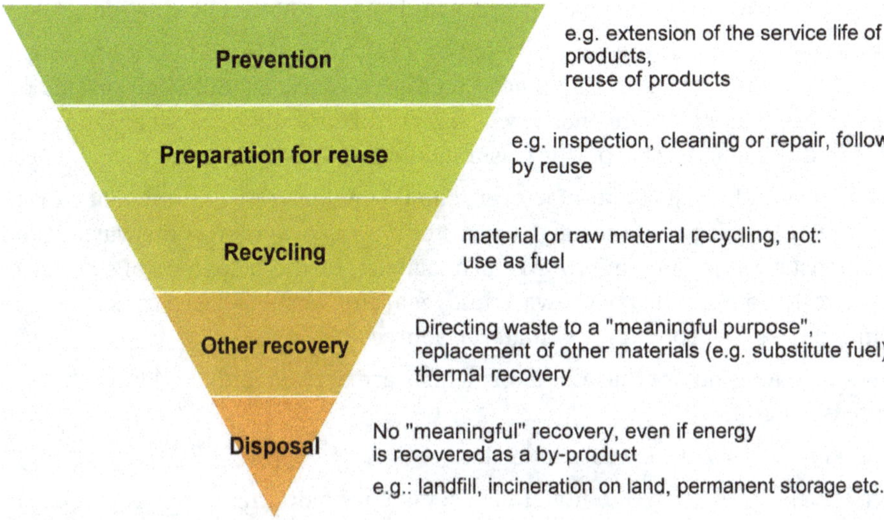

Figure 2.4 waste hierarchy of the Waste framework directive

Member states must promote measures that achieve a higher priority in the ranking. The amendment added a paragraph calling on member states to create incentives for applying the waste hierarchy and to use economic instruments or measures for this purpose. Examples include: polluter-pays fee systems, deposit systems, tax measures for products that can be reused or recycled, extended producer responsibility, or public awareness campaigns.

Another important foundation for future legislation is the extended producer responsibility also established in this directive. Member states may impose this responsibility on the respective manufacturers for certain waste streams or products. This includes the obligation to take back and properly dispose of products, to bear the costs of waste management for these products, and the obligation to make publicly available information on the reusability or recycling of products placed on the market.

In Germany, the contents of the 2018 amendment to the directive were transposed into national law through the Circular Economy Act (Kreislaufwirtschaftsgesetz).

2.2.1.2 EU Plastics Strategy

With the EU Plastics Strategy ("A European Strategy for Plastics in a Circular Economy") of 16 January 2018, the focus in waste and circular economy was placed on plastics. The strategy addresses the challenges prevailing worldwide and in the EU, including:

- Plastic waste and microplastics in the environment (land and sea)
- Inadequate collection for recycling in the EU (<30 %)
- Inadequate demand for plastic recyclates in the EU
- High CO_2emissions from thermal recovery and virgin material production

As a counterpart to this, it presents a vision of a circular-oriented plastics management. By 2030, all plastic packaging placed on the EU market should be reusable or recyclable, more than half of the plastic waste generated should be recycled, and sorting and recycling capacities for plastic waste should be quadrupled compared to 2015.

Both consumers and product manufacturers should demonstrate natural behavior in circular-oriented management through a structural transformation of the waste management industry (new companies for return logistics, further suppliers of standardized recyclates) and through a new awareness regarding circular-based consumer behavior and the use of secondary raw material sources.

To achieve such a vision, measures and principles are presented that address the following areas of action:

- Improvement of separate waste collection
- Improvement of the recyclability of packaging, incorporating producer responsibility and economic incentives
- Improvement of the economic framework conditions and the quality of plastic recyclates
- Deepening of knowledge on the circular management of plastics
- Reduction of plastic waste entering the environment
- Creation of a legal framework for biodegradable plastics

The EU Plastics Strategy combines an analysis of the current situation with the vision of implementing the circular economy for plastics in an ecologically and economically viable manner in the future. It thus also forms the basis for all further EU activities in this field.

2.2.1.3 Single-Use Plastics Directive

As a consequence of the presentation of the EU Plastics Strategy and to implement waste prevention, which occupies the highest position in the waste hierarchy, the "Directive on the reduction of the impact of certain plastic products on the environment" was adopted on 05 June 2019 by the European Parliament and the Council.

Based on study findings showing that single-use plastic items account for 50% of the plastic waste found on beaches, these are identified as a particularly major challenge. In addition, plastic fishing gear and oxo-degradable plastics are included in the policy measures of this directive. The aim of the directive is to reduce waste generation by promoting the reuse of products and the avoidance of single-use items.

These objectives are to be achieved through the following measures:

- Consumption reduction

  For beverage cups including their closures and lids, and for food packaging for food that can be consumed without further preparation, consumption is to be reduced across the EU. This is to be quantitatively assessed by member states in 2026 compared to 2022. The EU proposes the following measures for this purpose:

  - Offering reusable alternatives
  - Prohibition of free distribution to the end consumer
  - Adoption of marketing restrictions

- Prohibition of placing products on the market

  Member states are to prohibit the placing on the market of the following single-use plastic items since 3 July 2021, as well as items made from oxo-degradable plastics in general:

  - Cotton bud sticks, cutlery, plates, drinking straws, stirrers, balloon sticks
  - Food packaging made from expanded polystyrene for food that can be consumed directly without further preparation
  - Beverage containers made from expanded polystyrene

- Product design requirements

  A large proportion of the waste items found on beaches are beverage bottle caps. For this reason, the caps of beverage bottles are to be attached to the bottle. In addition, PET bottles are to consist of at least 25% recyclate from 2025 and 30% recyclate from 2030. This is to represent the average recyclate content of all PET bottles placed on the market in the respective country.

- Labelling requirements

  For sanitary pads, tampons, wet wipes, tobacco products and beverage cups, the packaging or the product must contain clearly legible and durable information about the available disposal options in the sense of the waste hierarchy. Furthermore, it must be declared that the item contains plastic and that improper disposal leads to environmental problems.

- Extended producer responsibility

  With reference to the Waste Framework Directive, extended producer responsibility applies to certain single-use items such as food packaging for food consumed without further preparation, bags and film packaging, beverage containers, etc. In particular, the manufacturers of these items bear the costs for awareness-raising measures (see below), for clean-up operations and, in part, for the collection of the waste and its further treatment.

- Separate collection

  By 2025, 77% of waste from PET bottles is to be collected separately. The reference quantity is the amount of PET bottles placed on the market in that year. By 2029, this share is to rise further to 90%.

- Awareness-raising measures

  These measures are to be understood as consumer information intended to encourage consumers to handle single-use items responsibly and to illustrate the impact of carelessly discarding waste.

2.2.1.4 Plastic Tax

At a special meeting of the European Council from 17–21 July 2020 on the "Recovery of the EU after the Covid-19 pandemic", measures were adopted to increase the EU's own resources. This included the levy on non-recycled plastic waste, which member states have been required to pay since 1 January 2021. The amount of the levy is 80 cents per kilogram of non-recycled plastic waste. In Germany, this levy has so far not been collected directly from industry, trade or consumers. However, manufacturers of single-use plastic products have been required since spring 2025 to make contributions to a single-use plastics fund administered by the Federal Environment Agency (see section 2.2.2.5).

2.2.1.5 The Packaging Regulation (Packaging and Packaging Waste Regulation – PPWR)

Regulation 2025/40 on packaging and packaging waste (PPWR – Packaging and Packaging Waste Regulation) replaces the packaging directive dating from 1994. It entered into force on 11 February 2025, with most requirements and obligations not applying until August 2026.

The PPWR includes further provisions for reducing packaging waste, increasing the recyclability and recycling capability of packaging, and for the separate collection of beverage containers made from plastic and metal. In addition, manufacturers must also declare conformity for packaging.

In addition to criteria for design for recycling at various performance levels (A, B and C), the PPWR lays the foundation for assessment methods for large-scale recycling. From the author's perspective, a significant new development is the tightening of requirements for the recyclate content of plastic packaging, which must amount to up to 35% from 2030 and up to 65% from 2040.

2.2.1.6 The Green Deal

The European Green Deal, presented in December 2019, is a strategy for the economy and the population of the EU with the aim of achieving net-zero greenhouse gas emissions by 2050. To this end, measures are intended to drive growth and the process of change through which the EU "should become a fair and prosperous society with a modern, resource-efficient and competitive economy" and "economic growth is decoupled from resource use." [EU19, p. 2].

In Figure 2.5 the key points and fields of action that comprise the Green Deal can be seen. Not only climate protection goals are pursued, but also the subject areas of mobility, food supply, the construction industry, the energy sector, ecosystems and the circular economy are addressed.

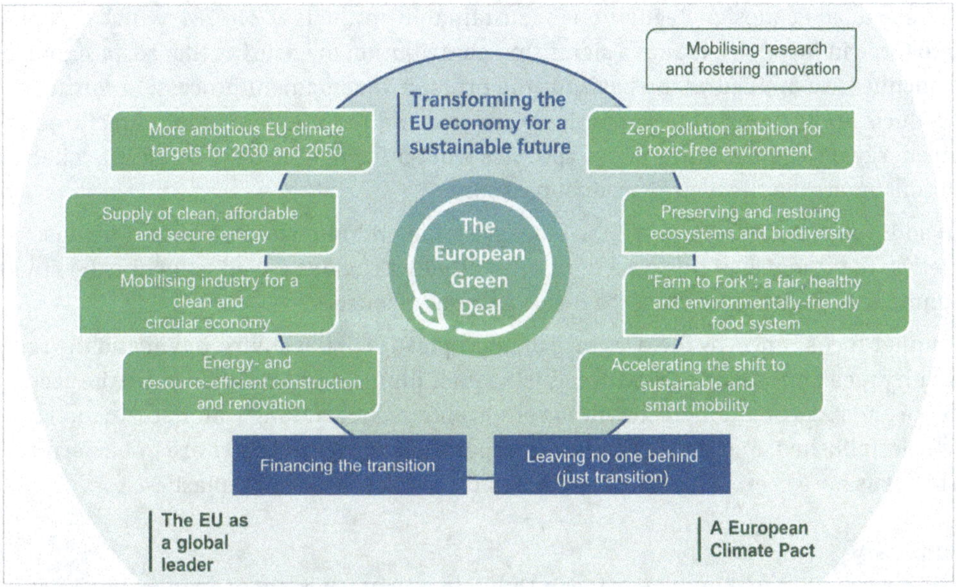

Figure 2.5 The European green deal [EU19, p. 4]

The far-reaching transformation process of the economy is to be implemented by the member states through political measures such as regulatory instruments, standards, investments, national reforms, etc. In particular, the importance of preserving ecosystems, the sustainable use of resources and the improvement of human health is to be significantly increased. This can help ensure that the social costs caused by environmental pollution are taken into account in product pricing through market-based, governmental instruments. This gives companies greater scope for action in terms of technological and product-related orientation compared to regulatory instruments (section 2.3.1).

All fields of action shown in Figure 2.5 are backed by concrete strategies and action plans that implement corresponding measures with a timetable. Under the field of action "Mobilising industry for a clean and circular economy", the Circular Economy

Action Plan was published in March 2020, addressing in particular the textile, construction, electronics and plastics sectors.

With regard to plastics, the EU Plastics Strategy of 2018 is to be consistently pursued further. By 2030, all packaging placed on the market in the EU is to be reusable or recyclable. The measures from the Single-Use Plastics Directive are supported by the action plan through a concretisation and harmonisation of the regulations. The PPWR comprehensively regulates the handling of plastic packaging and its recyclability as well as its minimum recyclate content. Since 2022, mandatory recyclate contents are to be introduced not only for packaging but also for the construction and automotive sectors. The so-called Ecodesign Regulation for sustainable products 2024/1781, which entered into force in July 2024, makes a significant contribution to ensuring that sustainability principles are applied as the basis for the product development process. Accordingly, products are to be durable, reusable, repairable and recyclable, and manufactured in an energy-efficient manner. It also specifies what information must be made available to consumers via the digital product passport.

In addition, a legal framework for biodegradable and bio-based plastics is being created so that the use of this type of plastic actually provides added value for the environment beyond the fact that it conserves fossil resources.

Another focus concerns the presence of microplastics in the environment and the regulatory measures associated with reducing their input. Scientific findings on the occurrence and risks of microplastics in the environment, in drinking water and in food are to be established, and measures for the separation of microplastics are to be derived. The focus here is on textiles and tire abrasion as sources of microplastics.

Summary

The directives issued in the European Union in recent years show an increasing tightening of regulations with regard to recycling quotas and recyclate content in products. Packaging producers will have to engage even more with improved, recyclable packaging design so that the material flow required for the demanded recyclate content remains available in the long term. The EU Packaging Regulation will accelerate this development significantly once more. The planned regulations regarding the use of recyclate in the construction and automotive industries should alert automotive manufacturers and, in particular, the supplier industry. Developments to establish new material flows for high-quality recyclates from engineering plastics are urgently needed in order to meet future demands.

2.2.2 Statutory Provisions

EU regulations are generally directly applicable in all member states. Directives, on the other hand, must be transposed into national law by the member states before they become binding for companies and consumers. In this context, the following section will address the five most important German laws for waste management and the circular economy:

- The Circular Economy Act (KrWG)
- The Packaging Act (VerpackG or VerpackDG)
- The Commercial Waste Ordinance (GewAbfV)
- The Single-Use Plastics Ban Ordinance (EWKVerbotsV)
- The Single-Use Plastics Fund Act (EWKFondsG)

2.2.2.1 The Circular Economy Act (KrWG)

This act serves to promote the circular economy in Germany and, with the amendment of the act that came into force on 29.10.2020, also represents the transposition of the Waste Framework Directive and individual provisions of the Single-Use Plastics Directive into German legislation. The aim is to ensure the protection of people and the environment in waste management.

The key elements can be summarized in the following non-exhaustive list:

- Introduction of the waste hierarchy (since 2018 also supplemented with concrete measures for incentivizing higher-priority recovery pathways)
- Provisions for establishing and promoting the circular economy
- Obligations and requirements for waste disposal
- Public-law waste management
- Regulations on the engagement of third parties in fulfilling the respective obligations
- Product responsibility for achieving the objectives of the circular economy
- Planning responsibility (waste management plans, approval of facilities, etc.)
- Waste advisory obligation
- Monitoring and documentation obligations

The majority of the provisions form the legal basis for further laws and ordinances. In addition, however, specific targets for promoting recycling are also established. Thus, the proportion of municipal solid waste prepared for reuse or recycled is to increase in three stages (an increase of five percentage points every five years), starting from January 2020 at 50 percent by weight, rising to 65 percent by weight by January 2035 [KrWG, § 14]. Furthermore, it is stipulated that public authorities ("federal authorities and legal entities under public law subordinate to the federal government", KrWG, § 45 (1)) must give preference in procurement to products that have been manufactured using recyclates and are characterized by recyclability, provided that "no unreasonable additional costs arise" (KrWG, § 45 (2)).

Further sections of this act that are significant for product placers on the market are sections 23–25, which deal with product responsibility and were supplemented in the amendment by a new duty of care. This requires that the fitness for use of products must be maintained for their distribution and that they must not become waste.

In principle, product responsibility applies to all companies that develop, manufacture, process or distribute products. It encompasses, among other things, the following points:

- Placing on the market of products that are reusable, technically durable and suitable for harmless and high-quality recovery after use
- Preferential use of recoverable waste or secondary raw materials
- Labeling of products containing hazardous substances
- Information on return, reuse and recovery options or obligations and deposit schemes through product labeling
- Take-back of products and the waste arising from them, as well as their proper recovery
- Assumption of financial and/or organizational responsibility for waste management
- Assumption of costs for clean-up operations by public authorities in connection with the products concerned

Product responsibility and the new duty of care are subject to a regulatory reservation pursuant to § 23 paragraph (4), i.e., they only apply after an ordinance has been issued by the federal government with the consent of the Bundesrat for specific products.

2.2.2.2 The Commercial Waste Ordinance (GewAbfV)

The "Ordinance on the Management of Commercial Municipal Solid Waste and Certain Construction and Demolition Waste" of 18.04.2017 regulates the separate collection and treatment of commercial waste in accordance with the requirements of the Circular Economy Act. Accordingly, commercial enterprises as waste generators must, insofar as technically and economically reasonable, carry out separate collection for the following fractions:

- Paper, cardboard and cartonboard
- Glass
- Plastics
- Metals
- Wood
- Textiles
- Biowaste

These fractions are to be preferentially directed toward preparation for reuse or recycling. In contrast to private households, commercial enterprises must also maintain documentation on separate collection, on the waste disposal companies that accept the waste, and on the unreasonableness of separate collection (if applicable).

Furthermore, the ordinance also regulates separate collection for construction and demolition waste and, in general, the sorting and recycling quota of pre-treatment facilities fed with commercial waste.

2.2.2.3 The Packaging Act (VerpackG/VerpackDG)

The German Packaging Act (VerpackG) ("Act on the Placing on the Market, the Take-back and the High-quality Recovery of Packaging") replaced the previously existing Packaging Ordinance (VerpackV) on 1 January 2019. In the process, it adopted and expanded the regulations for the dual system regarding the treatment of packaging subject to system participation obligations. In addition, it establishes the requirements for product responsibility pursuant to § 23 of the Circular Economy Act specifically for packaging. In order to implement the requirements of the Single-Use Plastics Directive and the amended Waste Framework Directive, the updated Packaging Act came into force on 1 July 2021.

The regulations on the obligation to participate in a system will not be discussed further at this point. Instead, the requirements for packaging in general and for beverage packaging in particular, as laid down in the law, will be presented.

The general requirements can be summarized as follows:

- Packaging volume and packaging mass must be limited to the minimum required to ensure the function of the packaging
- Recycling, reuse or recovery of the packaging must be feasible
- Environmental impacts when implementing this end-of-life handling must be limited to a minimum
- Hazardous or harmful substances must be limited to a minimum
- Maximization of reusability and the proportion of secondary raw materials in the packaging mass
- Creation of incentives for the end consumer (e.g., €0.25 deposit on single-use bottles)
- Labeling of packaging materials according to the numbers and abbreviations defined in the law

For single-use plastic beverage bottles made of PET, the following recycled content requirements apply with the amendment of the Packaging Act under § 30a:

- From 1 January 2025: 25% recycled content
- From 1 January 2030: 30% recycled content

The requirements are also considered fulfilled for a manufacturer if they can demonstrate compliance for all beverage bottles placed on the market within a year in total.

The law also sets out requirements for the recovery of packaging, which relate in particular to packaging subject to system participation obligations. In principle, the aim is to prioritize the preparation of packaging for reuse or recycling. With regard to plastics, the systems are obliged to recover 90% of this material in the participating packaging. Of this proportion, 70% has been required to undergo material recycling since 1 January 2022. This corresponds to a share of 63% of the total quantity of plastics to be recovered.

The VerpackG is to be replaced by the Packaging Law Implementation Act (VerpackDG) in order to align the legislation with the EU requirements of the PPWR. It includes a stricter recycling quota for plastic packaging from 2028 onwards.

2.2.2.4 The Single-Use Plastics Ban Regulation (EWKVerbotsV)

To implement the EU Single-Use Plastics Directive, the "Regulation on the prohibition of placing certain single-use plastic products and products made of oxo-degradable plastic on the market" was adopted on 24 June 2020 and entered into force on 3 July 2021. The regulation explicitly points out that the prohibition of products does not have to apply exclusively to packaging as defined under the German Packaging Act.

According to the regulation, the following products may no longer be placed on the market as single-use plastic articles since 3 July 2021:

- Cotton buds (with some exceptions)
- Plates, cutlery, stirrers and drinking straws
- Balloon sticks
- Food containers and beverage containers made of expanded polystyrene whose contents can be consumed without further preparation
- Products made of oxo-degradable plastics

These requirements are intended to reduce the entry of such products into the environment through careless littering and to reduce plastic consumption in general. In addition, the aim is to prevent oxo-degradable plastics, which break down into fragments through UV light, heat or oxygen, from remaining in the environment as microplastics.

2.2.2.5 The Single-Use Plastics Fund Act (EWKFondsG)

To implement the extended producer responsibility from the EU Single-Use Plastics Directive, the Act on the Single-Use Plastics Fund was adopted by the Bundestag on 11 May 2023. Its aim is to avoid or mitigate the negative impact of single-use plastic products on the environment and human health. A central component of the act is the establishment of a fund, the administration of which is the responsibility of the Federal Environment Agency. The revenues of the fund are intended to cover essentially the costs of collecting waste from single-use plastic products in heavily littered locations, for clean-up operations and for awareness-raising measures.

Manufacturers of single-use plastic products are obliged to pay a single-use plastics levy to reimburse the above-mentioned costs in accordance with the assessment notice issued by the Federal Environment Agency. This is calculated based on the quantity of products placed on the market by the manufacturers and the levy rates set by the Federal Ministry for the Environment, Nature Conservation, Nuclear Safety and Consumer Protection according to product type. Prior to this, manufacturers must register electronically and report annually the quantities of products placed on the market in the previous year.

The annual single-use plastics levy was collected for the first time in spring 2025. The registration obligation has been in effect since 1 January 2024.

2.3 Actors Along the Value Chain

As shown in Section 2.2, the European Union is pursuing the goal, within the framework of the European Green Deal, of transforming the traditional linear management approach into a circular economy-oriented approach. The protection of the environment and the conservation of natural resources are to be mentioned as key drivers of this development. The adopted and planned EU regulations are leading to a steadily growing number of statutory provisions in Germany.

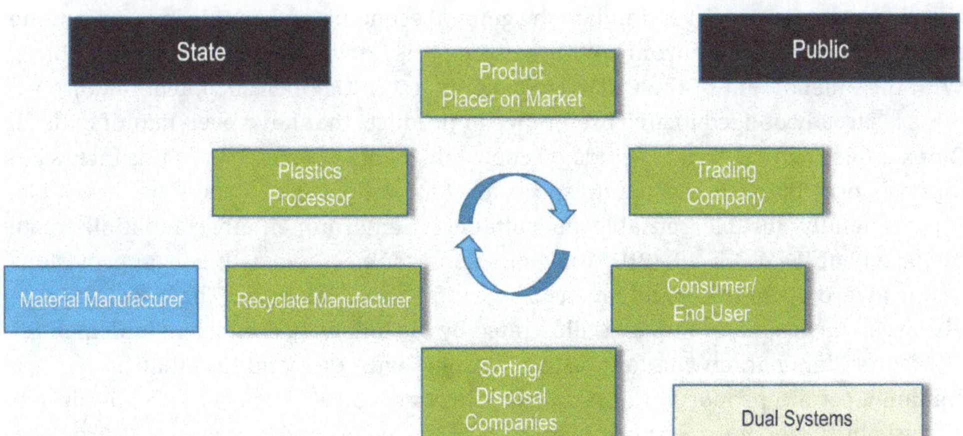

Figure 2.6 Actors along the value chain using the example of consumer products with packaging subject to system participation obligations

In Section 2.3, in addition to the state, the further actors along the value chain for an example product group are to be presented. The respective role, interests, possibilities for implementing those interests, dependencies and the significance of the actors are to be outlined in broad terms. In order to include the recycling facilities already established in Germany in the analysis, the focus will be primarily on consumer goods whose packaging must be certified through the dual systems. Products may include, for example, shampoo, shower gel, detergent or food packaged in plastic.

Each of the actors assumes its role along the value chain and, by virtue of this role, brings its own interests. Thus, it is the inherent interest of a product manufacturer placing goods on the market to generate profits.

In the following analysis, however, it is assumed on the other hand that none of the actors actively pursues an interest in contributing to environmental pollution and wasteful use of resources. With this in mind, the actors depicted in Figure 2.6 are presented one by one.

2.3.1 The State

Even though the state is not directly involved in the value creation process, it nonetheless exerts strong influence on events not only through its climate and environmental policy. In addition to its nationwide legislative function, the public sector also assumes further roles. It acts in part simultaneously as a purchaser, as a provider of waste disposal services and in the capacity of state supervisory authority.

As a purchaser or consumer, the state participates directly in economic activity. Apart from the possibilities available to the state to pursue demand-oriented economic policy approaches in order to stimulate the general economy and enable investment, the state can incorporate environmental policy aspects into the procurement of products. With the amendment of the Circular Economy Act in October 2020, federal authorities are already obliged to give preference to products that have been manufactured, for example, using recyclates or from renewable raw materials, even if this incurs additional costs. Further important aspects are that the products should be recyclable, repair-friendly, durable, reusable and suitable for environmentally compatible waste management [KrWG, § 45, (2)]. This obligation is subject to existing legal provisions on the award of contracts and on ensuring sufficient competition. That this legal provision can have a major impact is illustrated by the following scenario: municipalities put out tenders nationwide for waste containers with the condition that these must contain a certain proportion of recyclate. This has a direct impact on the producers of waste bins, who must first completely restructure their raw material procurement. This is followed by process-related modifications to the machinery, material and component testing, and close-meshed incoming and outgoing batch inspection due to the batch variations that the use of recyclates often entails. The product waste bin brings with it corresponding quantities that have a direct influence on the market for secondary raw materials.

As a provider of waste disposal services, it is the task of municipalities to ensure waste disposal for private households. However, this task is frequently delegated to private waste disposal companies as a service, which is regulated through corresponding contracts. In this respect, there is a relationship of dependency between the municipality and the private waste disposal companies, and at the same time a competition between both parties, since municipalities also partly carry out this task themselves.

State supervision of compliance with quotas, volume flows, or limit values is the responsibility of the relevant authorities. With regard to waste management, this is regulated in the Circular Economy Act (KrWG §§ 47–55).

In the following, the regulatory options available to the state within the framework of environmental and climate policy will be presented. With regard to the environmental policy objectives pursued therein, reference is made to Section 2.2.2.1.

Insert: the Environment As a Resource

First, a brief consideration of the environment is presented, which from an economic perspective provides various resources for humans, but also represents their habitat and that of animals and plants. The preservation of this habitat has gained considerable importance in the wake of evident environmental problems. According to Michaelis, the environment provides four important basic economic functions [Mic96, p. 5]:

- Raw material source as input for production processes
- Land and soil as a location for companies
- Land, sea, and air for the absorption of waste materials and emissions (pollutants, noise, and waste heat)
- Environmental goods accessible to the general public: breathing air, drinking water, landscape, recreational values

The use of these basic functions leads to mutual dependencies. When raw materials are extracted from the environment and products are manufactured from them, emissions and waste are generated during production. In addition, waste is generated through consumption. At moderate levels, these are absorbed by the environmental media of soil, water, and air, and distributed evenly through diffusion processes and partially neutralized (natural regeneration). If pollutants cannot be neutralized, they accumulate into stocks of pollutants that impair the generally accessible environmental goods. Industrialization, global population growth, and the increase in general prosperity, particularly in the western world, have led to the seemingly infinitely available, cost-free natural environmental goods (e.g., clean air) becoming increasingly scarce. From an economic perspective, Michaelis describes the environmental problem as a scarcity problem of environmental resources, the use of which must consequently be allocated among the actors [Mic96, pp. 6–7, Tis94, p. 3].

Overall, with a constant state of technology, an optimization problem exists between the use of the environment as a recipient of pollutants and the production volume of goods. Simply put: an unlimited production volume leads to continuously growing damage to environmental goods. Complete avoidance of environmental damage, on the other hand, leads to a standstill in the production of goods. Through monetary valuation of both factors, a damage function $S(E)$ as a function of the emission quantity E and a cost avoidance function $K(E)$ can be established, as shown in Figure 2.7. The variable S describes the opportunity costs for the resulting environmental damage and the variable K the opportunity costs for reducing the production volume in order to avoid emissions. The sum of both types of opportunity costs, related to the emission quantity, shows as a minimum the social optimum of the emission level E'.

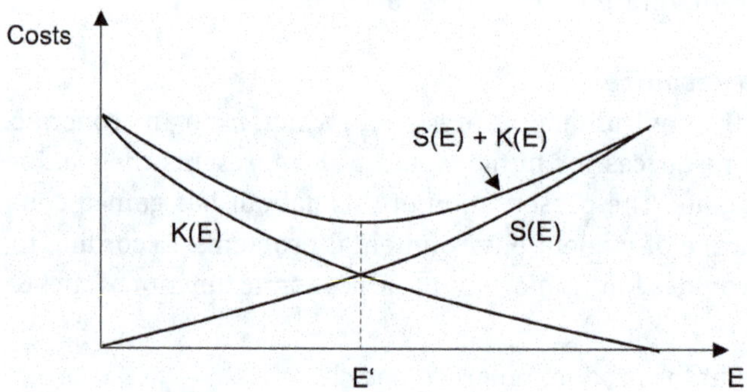

Figure 2.7 Damage function S and cost avoidance function K as a function of the emission quantity E, according to [Mic96, p. 10]

Even though the diagram fundamentally shows the interdependencies and the theoretical determination of the optimum, it is merely a model representation. The monetary valuation of environmental damage alone presents a major challenge. In addition, the functions are subject to the influence of technological change in the individual economic sectors, which entails a complex assessment of cost avoidance.

The allocation of these opportunity costs to the actors is the central question that environmental economics has been dealing with for decades. The aim is to guide the actors through various mechanisms so that they achieve the social optimum of emissions. There are several approaches to this, which are presented in overview in Figure 2.8 and will now be introduced in their basic features below.

Market-Based Instruments

In a market economy, the market — that is, the interplay of supply and demand — is regulated through price. This mechanism, which is important and effective for a national economy, does not, however, take into account the social costs caused by excessive exploitation of the environment. This represents a market failure, since the resources of the environment are not correctly valued.

Market-based instruments do not intervene in the market price mechanism, but artificially influence the resource price. This is done in two different ways [Ado08, p. 5]:

- Quantity control
  - The state sets a maximum permissible environmental load and issues certificates for this purpose that can be freely traded. The effect on the price results from the trading market.
- Price influence
  - Through environmental levies, the state sets the price for offsetting the environmental load, and the emission quantity results from the effect that the increased price exerts on production output.

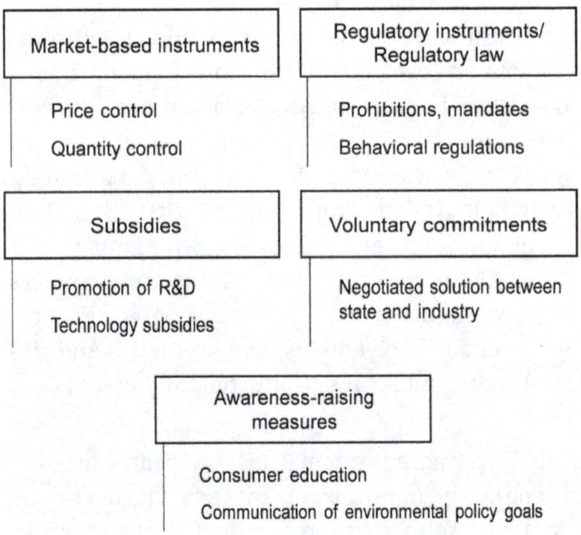

Figure 2.8 Overview of environmental policy instruments

The advantage of market-based instruments is that they utilize the market mechanism rather than undermining it through prohibitions. Economic actors thus retain flexible options for action, because they can take their own initiative to offset the increased prices through innovations and investments in research and development. Entrepreneurs become aware of the artificially induced scarcity of the environmental resource and this influences their entrepreneurial actions. They seek the most efficient way to avoid excessively high production costs in the long term. From the state's perspective, the key is to keep certificates scarce so that entrepreneurs are motivated to invest in future-oriented and more environmentally friendly technologies. In addition to the general limitation of environmental impact, the state can in this way ensure that entrepreneurs pursue environmental policy goals through innovations [RRR+08, p. 75 f.].

The plastic tax (see section 2.2.1.4) is an example of an instrument that influences the quantity of emissions through price. The purpose of this tax is to generate additional revenue to support member states following the COVID-19 pandemic. Nevertheless, this tax has the potential to acquire a steering function if it is passed on to companies or if the state itself more actively advances statutory provisions that lead to an increase in the recycling rate.

The pricing of greenhouse gas emissions is an example of a market-based instrument that operates through quantity control. At the third Conference of the Parties to the Framework Convention on Climate Change, the Kyoto Protocol was adopted in 1997, which for the first time contains international commitments to limit emissions. One of the instruments is international emissions trading, in which each country receives a certain quantity of emission certificates. If a country exceeds its national emission reduction targets, it can sell surplus certificates to other countries. To implement emissions trading in accordance with the Kyoto Protocol, the EU introduced the European Emissions Trading System (EU-ETS) in 2005 for the energy sector and for energy-intensive industries (steel mills, refineries and cement plants), and since 2012 also for air traffic within Europe (ETS 1). In Germany, the regulations on emissions trading were implemented in the Greenhouse Gas Emissions Trading Act (TEHG). On 20 July 2017, an amendment to the TEHG came into force, according to which plants with a production capacity of 100 t per day for the manufacture of the following polymers are included in the European emissions trading scheme: polyethylene, polypropylene, polystyrene, polyvinyl chloride, polycarbonates, polyamides, polyurethanes and silicones. In addition, some basic chemicals required for the production of polymers are also taken into account.

In 2021, a supplementary national emissions trading system (nEHS) was introduced in Germany, covering the areas of heat generation and transport that are not covered by the European system. The national system also follows the so-called “cap and trade” approach, in which a political upper limit on greenhouse gas emissions is set for all emitters in total (“cap”). The entities placing fuels on the market pay a price for every tonne of CO_2emitted into the atmosphere through subsequent combustion, which is settled through the purchase of certificates. This process is therefore also frequently referred to as “CO_2pricing”. The certificates are subject to free trading (“trade”), but move within a defined price corridor of between 55 and 65 € per tonne of CO_2.

In contrast to the European EU-ETS 1 system, in which emissions are generated directly at the plant operators and are also settled on this basis, the nEHS also obliges fuel producers to purchase certificates, but on the basis of the quantity of greenhouse gases subsequently emitted by industry or the consumer. The entities placing fuels on the market then pass on the CO_2price to industry and the consumer. The difference in the settlement system is due to the number of emitters present in the various sectors. Every motor vehicle in the transport sector and every gas heating system in residential buildings in the heating sector is an emitter. To avoid requiring this large number of emitters to participate in emissions trading, the entities placing the respective fuels on the market (petrol, diesel, heating oil, natural gas, liquefied petroleum gas, biomass and coal) have been obligated to do so. In 2028, the new European emissions trading system EU-ETS 2, which will then also cover the sectors of energy, manufacturing, construction, transport, building heating as well as agriculture and forestry, will replace the national emissions trading system nEHS.

Market-based instruments cannot, however, fully resolve ecological problems, because not all environmental damages can be internalized as external effects [Tis94, p. 55]. The accumulation of waste in the oceans or on coastlines is an environmental problem whose remediation incurs high costs and is limited to only a few product groups. Regulatory instruments are better suited here because they can be applied more directly.

Regulatory Instruments/regulatory Law

In addition to market-based instruments, the state can establish binding codes of conduct and sanction deviating behavior. This is probably the oldest instrument of environmental policy and was long regarded as the most effective means [RRR+08, p. 87]. Through clear requirements or prohibitions, environmental policy goals can be achieved with a high degree of certainty. While market-based instruments incorporate external costs into price formation and in this way steer companies in the direction that is correct from an environmental policy perspective, prohibitions or environmental requirements are capable of addressing specific grievances.

However, companies are restricted in their freedom by such regulatory instruments to determine their product portfolio, the use of raw materials and the choice of technologies themselves. Furthermore, competitive distortions may arise when product bans are imposed that competitors abroad are not subject to. The state faces not inconsiderable costs for monitoring compliance with the regulations.

From an economic perspective, the environmental factor is used free of charge in the manufacture of products, because there are no price-based limits that signal the scarcity of resources to the company. Regulatory instruments can only provide incentives for the use of new technologies to the extent that they are needed to meet required emission reductions or lower raw material consumption. If, however, these are dynamically adapted to the current state of technologies, the innovation impact can be assessed considerably more positively [RRR+09, p. 83].

The following examples of regulatory instruments can be cited in relation to the plastics industry:

- Specification of a recyclate content for PET beverage bottles
- Ban on certain single-use plastic articles

- Ban on the use of oxo-degradable plastics
- Specification of the recycling share of plastic waste
- Obligation to license packaging subject to system participation requirements

With the aforementioned regulations, the specific environmental goals can be achieved with a high degree of precision. Regulatory law also serves directly to avert hazards, as the following example also illustrates.

Example: Numerous single-use plastic articles are found on beaches and in the oceans, often carried there via rivers. Market-based instruments are not capable of specifically remedying this grievance depending on the degree of emission without also affecting other industries. Through the prohibition of some of these articles, however, this problem can be addressed directly.

The above-mentioned requirements affect in particular the entities placing products on the market, the raw material supply chain and the waste management industry.

Voluntary Commitments

Alongside regulatory law, there is by comparison a "softer" instrument that, strictly speaking, does not represent a state-introduced mechanism. Rather, it involves a unilateral, legally non-binding commitment made by an industry association upon itself. With regard to environmental issues, it concerns the achievement of an environmental policy goal within a defined period of time. The voluntary commitment of industry arises, however, following negotiations with the state, which result in mutually non-binding pledges. The state contributes specific environmental policy goals, to the fulfillment of which the economic or industry associations commit themselves after negotiation. In return, the state refrains from introducing binding environmental policy measures, such as requirements or taxes [Söl02, Tis94, pp. 260–263].

In environmental economics, voluntary commitments are viewed critically because the state has no means of enforcing the agreed goals through sovereign measures. Furthermore, the environmental policy goals are diluted in that the initially set targets are weakened through negotiation and, if applicable, only a small proportion of emitters are subject to the commitment. Once the goals have been achieved, there are also no further incentives to exceed them. It must also be assumed that voluntary commitments do not capture all external effects, since they do not apply to the economy as a whole but only to individual sectors. From a macroeconomic perspective, voluntary commitments are therefore to be assessed as inefficient according to Eickhof [Eic03, p. 6]. A further point of criticism refers to competition-threatening agreements between companies, which must be examined on a case-by-case basis. Söllner considers it problematic that the factor of "environmental protection" could be neutralized in competition with other companies [Eic03, p. 7, Söl02, p. 480].

For the state, voluntary commitments have the advantage that they can bypass complex legislative procedures and, depending on the subject of negotiation, produce practicable solutions. The costs of monitoring compliance with regulatory requirements are eliminated. Advantageous for the participating companies are potentially less ambitious targets and the opportunity to demonstrate their own environmental protection efforts to the public. Furthermore, companies are free to choose the means by which the agreed environmental goals are to be achieved. They can therefore act differently depending on the industry sector and pursue the goals in the most economically favorable manner.

According to Eickhof, voluntary commitments should only be pursued when it can be expected that they are superior to traditional methods in solving the problem [Eic03, p. 20]. Furthermore, the following criteria should be met [Eic03, pp. 20–21, Can01, p. 390]:

- Quantification of target specifications as precisely as possible in terms of fulfillment rate and deadline
- Addressing the differences of interest among the participating companies within the industry association and finding a consensus for the self-imposed commitments
- Inclusion of a large number of companies in the voluntary commitment
- Competition monitoring by the antitrust authorities
- Monitoring by an independent institution
- Credible threat of sanctions by the state in the event of non-fulfillment of the targets

The structure of the dual system in 1990 is an example of a voluntary commitment by industry. When the Packaging Ordinance came into force in June 1991, manufacturers and distributors were obliged to take back packaging and dispose of it. Industry was able to exempt itself from this obligation through a privately organised take-back and disposal system (the dual system). However, it had to meet the collection and recycling quotas stipulated by the Packaging Ordinance. What was once voluntary participation in a dual system, accepted by the state and incorporated into the Packaging Ordinance, has today become, under the Packaging Act, a mandatory requirement to participate in one of the dual systems.

In the course of establishing the Dual System Germany, the participating associations also entered into a further voluntary commitment. This concerned compliance with a reusable quota for beverage packaging. This commitment ultimately failed due to non-concrete targets, inadequate monitoring and the fundamental voluntary nature for individual companies [Can01, p. 390].

In 2016, the German Retail Federation (HDE) entered into a voluntary commitment with the Federal Ministry for the Environment in order to avoid a general ban on plastic bags. It committed itself and its members (over 360 retail companies) to no longer hand out plastic bags free of charge to customers. From a former 68 plastic bags per capita per year, consumption fell to approximately 18 bags per capita by 2021 [NN02-ol]. The EU plastic bags directive of 2015 envisaged a reduction in consumption to 40 bags per capita by 2025. Despite the significant decline in bag consumption, the Federal

Government decided, with the amended Packaging Act, to introduce a ban on lightweight plastic bags (15–50 μm), which has been in effect since 1 January 2022. This subset of plastic bags as a whole accounts for less than 1% of plastic waste in Germany [MHL20].

The Federal Ministry for the Environment launched the so-called "Cosmetics Dialogue" in 2015, in which a large proportion of cosmetics manufacturers participated. These manufacturers committed to voluntarily refraining from using microplastics in cosmetic products that do not remain on the skin (cleaning products for skin and hair) from 2020 onwards. Although the cosmetics industry states that it used 97.6% fewer microplastic particles in cosmetic products between 2012 and 2017 [NN03-ol], EU Regulation 2023/2055 as an amendment to the REACH Regulation has made the microplastics ban applicable to the aforementioned products as well, with a certain transitional period.

Subsidies

To implement environmental policy goals, subsidies are also used as a means, which also have an influence on prices. However, in doing so they do not make environmentally harmful activities more expensive, but rather make environmentally friendly activities cheaper [Kai13, p. 60]. As a result, the social costs are not attributed to the party responsible, but are distributed across the general public. If subsidies are measured directly against avoided emissions, external costs are internalized in this way, as was shown with the market-based instruments [Ado13, p. 328]. Subsidies place a long-term burden on the state budget and can often not be easily dismantled due to the planning security granted to companies.

However, funding measures are frequently used, for example, to further develop promising technologies that are considered too risky by companies or investors, or that require a long lead time before they can be operated economically. One example of this is the Renewable Energy Sources Act (EEG), which enables producers of renewable energies to be competitive through corresponding remuneration rates. According to Sprenger and Rave, however, this does not formally constitute a subsidy, because the additional costs are passed on to households. Nevertheless, this measure has the character of a subsidy due to the preferential treatment of the technology through price control [RS03, p. 114].

Otherwise, there are numerous funding opportunities in the area of research and development through EU, federal or state funds in the form of innovation networks, individual grants or advisory funding. These funding programs in the areas of environmental protection, circular economy and resource conservation have increased significantly in the past. In this way, industries from these sectors can grow and play a larger role as market participants.

Awareness-Raising Measures/promotion of Environmental Awareness
Through education and central information on the environmental impact of consumer behavior and the use of raw materials and products, the state can influence consumers and the economy. It formulates its environmental policy goals and identifies measures that businesses and consumers can take to reduce environmental impacts. This can raise awareness among actors for a more sustainable use of available resources. For this purpose, strategy papers, programs and campaigns are launched.

The Federal Ministry for the Environment has published several such measures in recent years, dealing with waste prevention, increased recycling rates and the stronger expansion of reusable packaging. The "5-point plan for less plastic and more recycling" and the waste prevention programme are two of these initiatives that programmatically demonstrate how the state intends to guide the relevant actors (industry, trade and consumers) towards greater environmental awareness.

The effect of such measures requires that a large proportion of the actors can comprehend the information and the resulting measures. The formation of state will, and thus also the environmental policy orientation, is the result of numerous factors. The publicly prevailing opinion, influenced by international, sometimes radically conducted environmental movements on the one hand and lobby associations from the business environment on the other hand, can lead to a distorted picture in the area of environmental protection, which insufficiently takes scientific findings into account.

2.3.2 The Dual System

As already explained in Section 2.3.1, the dual system was created as a self-commitment by industry in the course of the introduction of the Packaging Ordinance in 1991. In the following, the historical context of the dual systems will not be examined in greater detail; instead, the principle applied today and the role as an actor along the value chain will be presented. In contrast to municipal, public-law waste disposal, which is responsible in the broadest sense for residual and organic waste, it is the task of the so-called "systems" to ensure the collection, sorting and processing of packaging subject to system participation requirements. The term "dual" arose from the addition of private-sector waste disposal by the systems to the existing municipal waste disposal. The extended producer responsibility of the Circular Economy Act (KrWG) is regulated for packaging in the Packaging Act (VerpackG or VerpackDG). Accordingly, first placers on the market must ensure, prior to placing products on the market, that the packaging is taken back across the board and recycled in accordance with the legislation. This is achieved by participating in one of the dual systems. Specifically, the placer on the market (producer, trading company, importer, online retailer) concludes a contract with one of the systems, in which the latter is commissioned with the collection, sorting and recycling of its packaging. In return, the dual system releases the placer on the market from its obligation under extended producer responsibility.

In addition to system participation, producers placing packaging on the market are obliged to register with the Central Agency Packaging Register and to publicly deposit data on the quantity of packaging placed on the market.

Manufacturers and first distributors must have each packaging licensed by a system. Frequently, a basic fee and a quantity-dependent licensing fee are charged for the packaging, which is categorized according to the material of the packaging. With these revenues, the dual system commissions the collection, sorting and processing in accordance with the statutory recycling quotas per material. The commissioning is sometimes preceded by tenders, to which waste disposal companies can submit bids. The licensing revenues are thus offset by the fee-financed commissioning of the recycling obligation. In most cases, the dual systems themselves also take on a portion of the recycling, for example by producing recyclate from the lightweight packaging collected from the recyclables bin. The system generates additional revenues from the sale of the recyclates.

The dual systems compete with each other, so the level of licensing fees varies depending on the provider and type of material. In order to meet the nationwide collection requirements and the recycling quotas, they must conclude contracts with the partners needed to fulfil the statutory obligations, which place a burden on the expenditure side. This then results in the minimum level of licensing fees for those placing products on the market.

2.3.3 The Public

The publicly prevailing opinion is a phenomenon described by many political theorists, philosophers, historians and sociologists, whose definition is difficult to grasp. Historically, moral concepts and social pressure to conform are derived from it. In the course of the Enlightenment, the term increasingly became a socio-political category. According to Reese-Schäfer, who extensively presents the theories of Jürgen Habermas on the structural transformation of the concept of the public sphere, the public sphere is a "term for describing the communicative space between civil private life and the state" [Ree06, S, 93]. Public opinion can be understood as a collective product of communication in the public space. It must be distinguished from the individual opinion of a single person. According to Neidhardt, it is rather "the totality of opinions that are publicly communicated before a [...] audience." [Nei07, S. 27]. In this context, the opinion expressed publicly is influenced by various effects, such as the self-presentation of the actors, strategically calculated reactions among the recipients and the counter-speech to other publicly expressed opinions [Nei07, S. 27].

The media play an essential role in the articulation and formation of public opinion. They convey theses, arguments and opinions in public discourse, whereby the emphasis placed on topics and the selection of actors who express an opinion in public depend on the respective socio-political background of the editorial teams. This can lead to an emerging dominant opinion of the public being characterized by a one-sided view of a matter, because under certain circumstances essential actors do not participate in the formation of opinion in the public sphere. The idea that public opinion, according to Habermas, "is purified through critical discussion into true opinion" [Hab62, p. 118] is an idealized notion that presupposes that all citizens have free access to the public sphere and also make conscious and active use of it. In the classical media, journalists, politicians, scientific experts and celebrities account for a large proportion of the actors who shape the "public sphere." Added to these are actors in social networks, who by now find a large stage in the digital public sphere. Here a further problem of an apparently formed public opinion presents itself. Alongside the publicly dominant opinion that forms in the discourse of the real public sphere, there can simultaneously exist a divergent, publicly dominant opinion that forms in the digital public sphere. Only when the same actors take up theses, arguments and opinions from the respective other "world" can a merging discourse come about. The recipients who engage with public opinion must consequently also make use of access to both the analogue and digital public sphere.

Due to the challenges described in forming a publicly dominant opinion, a fatal development has occurred with regard to the image of plastics. Obviously, scientists or experts as actors have only insufficiently participated in the public discussion. This is the only way to explain how scientifically unsubstantiated theses in the public sphere have led to the perception that plastics are fundamentally harmful to the environment and should be replaced by other materials. Graphically presented images of the waste problem in the world's oceans, linked with slogans against plastics without factual analysis of the causes, have formed the focus of the opinions presented in the public sphere. In his book "The Plastics Paradox", Chris DeArmitt holds environmental groups, which advocate their theses without scientific evidence but with great passion, partly responsible for this development [Dea20, p. 183].

Dependency and Significance

The public as an actor is a quantity that emerges passively through public discourse. It depends on the composition of the actors who serve as mouthpieces and opinion-formers in the public sphere. The media plays an essential role in this, because it forms the stage for a large part of the theses expressed. Social networks have created their own channel for participation in public discourse for an audience that only partially overlaps with the analog public.

The importance of the publicly prevailing opinion is to be rated as high. Politicians adopt the publicly prevailing opinion for their own political agenda in order to be elected with their party as part of the government. Inevitably, legislation is also influenced by this. The informed citizen engages with public opinion on a subject area, derives their own opinion from it and decides in favour of a party through the act of voting. On the other hand, they are also themselves a consumer and are influenced by the public in their purchasing decisions. Confronted with the negative image of plastics prevailing in the public sphere, they may subconsciously opt for packaging with a (seemingly) positive environmental record.

Product manufacturers and material producers face an uphill battle against prevailing public opinion, because they are frequently accused of exclusively pursuing their own economic interests. Accusations of greenwashing quickly arise when producers develop sustainability concepts or carry out initiatives to promote the circular economy. In particular, manufacturers of virgin plastics find themselves exposed to strong public criticism, which has led to a negative image in the public eye.

2.3.4 The Consumer

The end consumer or consumer purchases food, beverages and other everyday products, for example from stationary retail stores. They have the free choice of which trading company they buy which products from which product placer on the market.

Interests and Their Implementation

Usually, the consumer wants to be able to choose high-quality products from a large selection of products, and at a reasonable price. Depending on socialization, purchasing behavior is also characterized by a fundamental awareness of sustainable and environmentally conscious action. If a corresponding range of products is available, the consumer exercises their demand power: what they do not want to buy, they simply do not buy.

Dependencies and Significance

Fundamentally, the consumer represents the market-economically decisive role as a demander and thus has a steering function. This function is exercised by the consumer in a sanctioning manner towards trade and the producer. These usually only learn after the manufacture of the products whether they are accepted or not.

A dependency of the consumer on the retail trade can be seen in the not unlimited product selection available. In addition, the consumer is influenced in his purchasing behavior from several sides. He is informed by the state about the sustainable handling of products and sensitized to an increased environmental awareness. The conformity pressure created by the public with regard to an apparently correct handling of plastics and their products is additionally a not negligible component that can shape the consumer. Another influencing factor concerns the marketing strategies of producers and retailers. This relates on the one hand to pure product advertising and on the other hand to the image cultivation of companies to act in a particularly sustainable and ecologically responsible manner. The former connects to the inherent interests of the consumer as a buyer, the latter is a reaction to the publicly prevailing opinion, for example with regard to negatively perceived plastic packaging.

2.3.5 The Product Placer on the Market

Product placers on the market are companies that place a product on the market for the first time. Producers of consumer goods (e.g. packaged food, beverages, cosmetic or cleaning products) fill their goods into packaging, which they either produce themselves or obtain from a plastics processor. They sell their goods to trading companies, which then offer them to consumers. For the sake of simplicity, the following analysis is limited to this distribution channel. Special forms of companies that simultaneously take on multiple roles along the value chain will be addressed separately.

Interests and Their Implementation

Producers pursue the goal of generating profit from the sale of their goods. They want to market their products as effectively as possible and strive for a positive public image. To this end, they develop products that are attractive to consumers and conduct marketing campaigns to highlight the advantages of the products and draw consumers' attention to themselves. In doing so, these companies rely on current trends and social movements that are conveyed in public perception and through the media. This gives rise to the sustainability concepts presented by many producers, which serve to obtain or preserve a positive image among consumers. Some publish the implementation of regulations already required by law, others publish industry-wide voluntary commitments concluded under pressure from the state, and yet others present a sustainability concept that goes beyond this.

Dependencies and Significance

Producers are obligated by the state to comply with legally prescribed regulations. This concerns requirements, prohibitions and established quotas, but also market-based instruments of the state. By law, they are also obligated to license their packaging with one of the dual systems and to register as a company with the Central Agency Packaging Register.

In most cases, producers distribute their consumer goods through retail, with which they negotiate the inclusion of products in the retail company's assortment. Here, brand manufacturers face increasing competition from retailers' own private labels or from innovative start-ups and so-called influencer products. In particular, food start-ups rely on so-called sustainable products with more environmentally friendly packaging [Lei19-ol]. The bargaining power of retail companies is to be assessed as strong due to the competitive situation vis-à-vis brand manufacturers, as some examples from the recent past show, in which there were temporary delistings of branded products at retailers such as Edeka, Rewe or Lidl [NN05-ol]. On the other hand, consumers' expectation of finding certain branded products in the assortment leads to a better negotiating position on the part of brand manufacturers in some product categories.

In addition to pressure from trading companies, producers are confronted with diffuse or concrete sustainability demands from the public. These also influence consumers, who in turn are the important end buyers of the products.

As shown in Figure 2.6, producers are dependent on important players such as trading companies. In this power structure, they must assert their position not only in terms of pricing, but increasingly also through their corporate profile in matters of environmental awareness. Consumers, influenced by prevailing public opinion, pay attention, for example, to so-called sustainable packaging solutions with a high proportion of recyclate. How seriously major producers such as Nestlé or Procter & Gamble take these wishes and demands is demonstrated by the large-scale sustainability programs of these corporations. Thus, Nestlé has, according to its own statements, saved 142,000 t of packaging material since 2015, aims to make all packaging recyclable or reusable by 2025, as well as to refrain from using single-use plastic items in its own consumption that are not recyclable, and is investing two billion Swiss francs worldwide in an "innovative circular economy". In addition, the company has founded "The Nestlé Institute of Packaging Sciences", which develops sustainable packaging [NN06-ol].

2.3.6 Trade

Also with regard to trade, the following will focus on the classic distribution channel for consumer goods (food, beverages, detergents, etc.) via stationary retail. The latter purchases goods from producers and offers them together with own-label products to the consumer as the end user.

Interests and Their Implementation

Trading companies pursue the goal of generating profit through the sale of goods. Depending on their orientation, they offer a broad product range in order to provide consumers with a wide selection of products and to meet their expectations. Retail responds to the environmentally conscious trend, which is perceived by the public and reinforced by numerous initiatives, with its own sustainability efforts. This in turn serves to maintain a positive image with the public and with individual consumers.

Dependencies and Significance

With regard to the state, the trade sector is not as strongly dependent as producers when it comes to environmental policy issues in the plastics sector. For private-label products, retailers bear the responsibility of the first placers on the market; for traded goods, this is handled by the producer.

The public is pushing for sustainable products with recyclable packaging. Aware that a negative public image can lead consumers to shop with a competitor instead, trading companies actively bring their environmental topics to the forefront. Ultimately, their concern is consumer satisfaction. Through marketing campaigns, however, they can also influence consumer purchasing behavior.

In the power structure with producers, different conditions emerge depending on the product category. For some product categories, there is a high brand affinity among consumers, which leads to a better negotiating position for the brand manufacturer; for other product categories, the purchasing power of the trade is greater [NN07]. In principle, the trade can determine its own product range and builds competition with suppliers through private-label products. The own products are sometimes cheaper than branded products, and in addition the retailer can determine the packaging philosophy itself and actively use it for advertising. For example, Aldi Nord and Aldi Süd together founded the initiative “ALDI Verpackungsmission” (ALDI Packaging Mission), which aims at the sustainable use of raw materials for packaging. By 2025, 30% of packaging material is to be saved for private-label products compared to 2015. By 2022, all private-label packaging is to be recyclable and also partly made from recyclate. In addition, Aldi supports the “Plastic Bank” initiative, which establishes collection points for plastic waste in less developed countries in order to prevent uncontrolled waste disposal [NN07].

2.3.7 The Sorting/disposal Company

The waste collected in the recyclables bin or the yellow bag reaches sorting plant operators via transfer stations, which also frequently operate their own disposal vehicles and handle household waste collection. In addition, there are numerous other disposal routes in the commercial sector that are served by these companies.

Interests and Their Implementation

First and foremost, sorting and disposal companies must fulfill the contracts entered into with the dual systems or municipalities. In order to utilize plant capacity, there is an interest in processing the largest possible quantities of waste with a high proportion of reusable materials (metal, glass, plastic, or paper). The sorted waste fractions are then sold freely on the market to recyclate manufacturers or affiliated companies. The sorting quality of the plants determines the quality of the output material, whereby the minimum sorting quality for the individual fractions is prescribed. The disposal companies are therefore concerned with achieving a sufficiently high sorting quality using the most up-to-date technology available.

Dependency and Significance

Plant operators must adapt to new regulations on the design of packaging with regard to individually sortable material streams, as these may lead to shifts in quantities toward mono-plastics and reduce the volume in the mixed plastics fraction. In addition, the operators of these plants must respond to new recovery and recycling quotas.

The dependency on the dual systems with regard to the awarding of contracts for collection and sorting is to be rated as high. On the other hand, the market for secondary raw materials is also still characterized by partially low demand, which is due to the low virgin material prices in recent years. A further challenge lies in the insufficient pre-sorting by the consumer. Many waste items end up in the mixed fraction, even though, for example, the removal of aluminum lids from yogurt containers by the consumer would have a significant impact here.

Overall, sorting and disposal companies have few opportunities to bring about positive changes, for example in the acceptance of secondary raw materials. The instrument to convince producers to use recyclable packaging or to increase recycled content is lacking.

2.3.8 The Recyclate Manufacturer

A manufacturer of recyclates is, in the simplest case, a company that purchases sorted fractions from a disposal company or reject material from plastics processors and produces recyclates from them.

Interests and Their Implementation

The company aims to generate profit through the sale of high-quality recyclate. In addition to the quality of the recyclate, the delivery quantity over a long period of time is decisive for the successful sale of the material. To this end, the recyclate manufacturer enters into long-term supply contracts in the area of post-industrial waste with plastics processors who deliver their reject material or residual materials to it. The certification of recyclates, which is required in many industries, is a further building block for offering recyclate on the market in the best possible way.

Dependencies and Significance

For recyclate manufacturers, competition with frequently cheaper virgin material presents a major challenge with regard to the lack of demand for recyclates. Recyclates are often associated with lower quality and higher prices compared to virgin material, and these are not exactly convincing sales arguments. The sustainability trend among producers, consumers, and the trade helps recyclate manufacturers in selling their goods. In some cases, producers are rather skeptical toward recyclate. Price pressure forces the plastics processor, who manufactures the packaging for the producer, in many cases to use virgin material. The fact that in this constellation there is no direct supply relationship between the recyclate manufacturer and the producer further limits the persuasive power of recyclate manufacturers vis-à-vis the producer. Plastics processors need plastic material of consistent quality over an extended period of time to fulfill their supply contracts. This frequently presents the recyclate manufacturer with major challenges.

Overall, recyclate manufacturers must convince with good and consistent quality combined with long-term delivery capability in order to sell their goods. The competitive situation with virgin material manufacturers shows that long-term, industry-independent success is only possible through the increase of recyclate quotas or other statutory provisions.

2.3.9 The Material Manufacturer

Material manufacturers refers to those companies that produce and sell virgin plastic from the raw material crude oil. Sales to plastics processors are usually made through distributors. The plastics used in the largest quantities in the packaging sector are polyethylene and polypropylene. The corresponding manufacturers of these plastics are large corporations in the chemical industry.

Interests and Their Implementation

These companies have an interest in generating profit through the sale of the plastics and additivated plastic compounds they produce. In doing so, they can manufacture their materials in large quantities with high quality using processes established over decades, and thus guarantee long-term volume commitments. In the course of the environmental debate, they frequently find themselves directly confronted with harsh criticism for being raw material suppliers for improperly disposed packaging. These corporations attempt to counter this negative image in various ways, by acquiring recyclate manufacturers or taking over recycling themselves, or by participating in internationally organized joint projects to promote the circular economy.

Dependencies and Significance

As described above, a public image of plastics is conveyed that associates them with fundamentally environmentally harmful properties. Not the way humans handle plastic waste, but the plastic itself is the problem — this is the impression that frequently arises. The pressure in society to change the grievances has on the one hand led to a multitude of statutory provisions and on the other hand to demands from producers and the trade for increased use of recyclates. Material manufacturers are confronted with this development, which can also already be seen when looking at the production and processing volumes of virgin material compared to recyclate. The volume of virgin material used in plastics processing in Germany fell by 10.2% in 2019 compared to 2017 and by a further 4.4% in 2021 compared to 2019. In the same periods, the recyclate volume increased by 2.5% (2019/2017) and 17.6% (2021/2019) respectively (with an overall decline in processing volume of 1.4%). In the production of virgin plastics, there was a decline of 6.6% in Germany in 2019 compared to 2017; however, in 2021 a growth of 4% compared to 2019 was achieved again [Con19, p. 52, 54, Con21, p. 55, 57]. The fact that 90% of the total volume of plastics processed worldwide still consists of virgin material, however, also demonstrates the still immensely large significance of material manufacturers to date. Due to the large-scale production of virgin material, it can often be offered at lower prices, and with lower fluctuations in material quality.

2.3.10 The Plastics Processor

As already mentioned several times above, the focus here should also be placed on processors that manufacture plastic packaging for consumer goods, such as food, beverages, cleaning agents, etc., for producers. The latter orders packaging from the plastics processor in the appropriate quantity and with the negotiated conditions and specifications. The plastics processor in turn obtains the plastic granulate from a material manufacturer or distributor, or a secondary raw material from a recyclate manufacturer.

Interests and Their Implementation

The processor generates profit through the sale of self-manufactured plastic packaging to product placers on the market. In doing so, it aligns itself with their requirements regarding material, desired properties, design, and price. The exclusion of recyclate or the requirement for a specific recyclate content is also stipulated. The plastics processor will fulfill these requirements with the lowest possible manufacturing costs in order to generate profit for itself. It will keep itself informed about current trends and accordingly build up know-how, for example in the processing of recyclate. Here, multi-component technologies, the response to material fluctuations through control engineering, and in particular a strong expansion of quality management for incoming raw materials and for manufactured products can play a role.

Dependencies and Significance

A strong dependency arises from the supply relationship with the product placer on the market, who wishes to keep the prices for packaging low. Here, plastics processors find themselves confronted with the situation of having to bring the frequently higher prices for recyclates into negotiations with the product placer on the market. The latter are reluctant to accept the higher prices compared to virgin material, but may nevertheless wish to follow the trend of increasing the recyclate content. Other producers accept pure inline recycling, which has already been implemented as standard for many years. Not only the use of recyclate is the subject of negotiations, but also recyclable design. The recyclability of packaging can be assessed using evaluation schemes from various institutes. Here, plastics processors must build up knowledge and experience relating to the sorting and recycling of plastics. In some cases, however, this also occurs at the producers, who wish to become active in this area themselves. The challenge for plastics processors therefore also consists in co-selling the recyclable end-of-life condition of their packaging. This will become increasingly important in the future.

The possibilities for plastics processors to assert their own interests are to be rated as rather limited. Ultimately, processors must follow the demands of producers, whereby larger packaging manufacturers certainly have greater opportunities to actively offer producers their own developed, potentially recyclable packaging solutions and to convince them with their price-performance ratio.

2.3.11 Bundling of Multiple Roles

The interrelationships shown so far each concern individual companies that occupy a single role along the value chain. The dependencies of these roles are still a lived reality, although there have been overlaps in many cases for a long time, which vary depending on the industry. For example, trading companies are sometimes simultaneously large placers of products on the market (private labels) and producers are simultaneously plastics processors who manufacture their own packaging. A further example concerns the disposal industry, where companies are present on the market that are simultaneously a dual system, sorting and disposal company, and recyclate manufacturer. By bundling multiple roles, dependencies can be reduced and the significance of these companies in the respective industry can be increased.

Particularly in the last five years, company acquisitions or mergers have been observed in the course of the recycling market moving more into focus. The two largest virgin material manufacturers of polyolefins, LyondellBasell and Borealis, have been increasingly entering the recycling market since around 2017. In 2018, LyondellBasell and the company Suez S.A., one of the world's leading waste disposal companies, commenced recycling activities in the jointly managed joint venture QCP (Quality Circular Polymers). Suez supplies the plastic waste as raw material and handles processing, while LyondellBasell is responsible for the distribution of the recyclates. Borealis acquired the German recyclers mtm plastics GmbH and mtm compact GmbH in 2016 and the Austrian recycling company Ecoplast in 2018. The chemical group Evonik is also increasingly expanding its recycling activities. By 2030, Evonik aims to achieve €350 million in revenue from the sale of secondary raw materials [NN08-ol], which accounts for 3% of total group revenue in 2020, and just under 11% with reference to the plastics division "Smart Materials". In this way, the virgin plastics material manufacturers are also serving the recyclate market, which can compensate in the long term for the presumably declining production volumes in the virgin material business.

The Schwarz Group, with 500,000 employees, is one of the world's largest trading companies and unites almost all roles along the value chain. The two retail companies Kaufland and Lidl are supplied with private-label products by numerous own production facilities, which operate under the umbrella brand Schwarz Produktion. For the PET material cycle, Schwarz Produktion operates three plastics and recycling plants within the beverage producer MEG, which recycle the deposit-bearing PET bottles back into new bottles. In this way, the Schwarz Group secures its own raw material repeatedly for its own purposes and is largely independent of external raw materials for this material stream. A further component of the Schwarz Group is the company PreZero, which operates as a disposal and recycling company in Europe and North America. The areas of activity of PreZero include waste collection, sorting and recovery, the production of recyclates from polypropylene, polyethylene, and polystyrene, and the "Dual" business segment. This means that PreZero has established its own dual system in which customers can have their packaging licensed. PreZero has held a majority stake in the KBG Group since September 2021, which produces granulates from LLDPE, LDPE, PP, and PS and processes them into films, packaging, and carrier bags [NN09]. In the packaging sector, the Schwarz Group ensures a raw material input stream for recyclates that is under its own responsibility. In this way, it directly influences the required delivery quantities and the required quality. The development of recyclable packaging can also be mapped in-house with the involvement of the know-how of the plastics processor, the producer, the sorting plant operator, and the recyclate manufacturer. For the Schwarz Group, the remaining actors are the public, the state, and the consumer, on whom it is inevitably dependent. However, the individual representatives within the corporate group can respond flexibly to the needs and demands of these actors.

2.4 Selected Material Streams and Their Characteristics

After in section 2.3 the individual actors along the value chain were presented with their interests, dependencies, and respective significance, the following section provides an overview of the five material streams with the largest processing volumes. Industry-specific characteristics are used as explanations for the respective recyclate quantities used and the recovery methods.

For all material streams, a distinction can be made in the end-of-life state between post-consumer and post-industrial waste. Post-consumer waste arises after use by the private or commercial end user. Post-industrial waste is generated during the manufacture or processing of plastics. In this case, this waste must have been handed over to a disposal company or recycling facility with a waste code number. Waste that is passed on without a waste code number or processed internally is referred to as a by-product (see § 4 KrWG). Figure 2.9 shows the temporal development of the plastic waste volume, broken down by post-consumer and post-industrial waste including by-products.

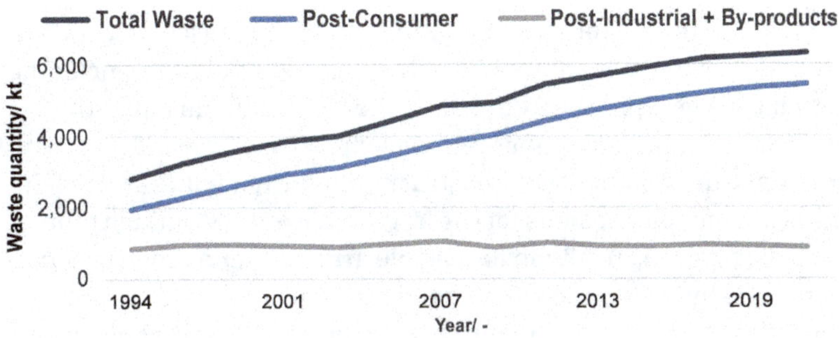

Figure 2.9 Temporal development of plastic waste volumes in Germany [Con21, p. 104]

The increase in total plastic waste volume is mainly attributable to the rise in post-consumer waste. The constant trend in waste volumes in the post-industrial sector can be attributed to two opposing effects. On the one hand, the increase in production volume over time leads to higher production waste. On the other hand, the improvement of processing methods has led to a reduction in reject rates. In general, post-industrial waste can be recovered well due to its cleanliness and purity of grade.

The material flows that will receive particular attention in the following are:

- Packaging
- Products for the construction industry
- Tired vehicles
- Electrical and electronic equipment
- Products for the agricultural industry

The post-consumer waste generated in these sectors is subject to different disposal and recovery routes, from which the main challenges can be derived. Packaging accounts for the largest share of total plastic waste, followed by construction products, electronic equipment, agricultural waste, and vehicles. Packaging has a short service life and typically becomes waste in the same year it is manufactured and used. Construction products, on the other hand, have a long service life and only become waste after a considerable time delay (> 40 years). This means that new developments in the recycling of recyclable products can be implemented more quickly in the packaging industry than is the case for products from the construction industry. A further fundamental challenge arises from inadequate waste separation by consumers and from multi-component products that cannot be separated, which predominantly results in heterogeneous waste streams. Material recycling is made more difficult as a result, since scalable processes for separating multi-component waste or products are frequently lacking. The material recycling of single-grade waste (e.g., PET bottles), by contrast, has been established for years.

According to the Conversio study "Stoffstrombild Kunststoffe in Deutschland" (Material Flow Analysis of Plastics in Germany), 14.04 million tonnes of plastics, including recyclate and by-products, were processed in 2021. After adjusting this quantity for import and export effects, 12.4 million tonnes remain, which were demanded or consumed in Germany as plastic products. The construction industry, with reference to the five material flows mentioned, accounts for the largest share at 3.36 million tonnes. This is followed by packaging at 3.20 million tonnes, electronic equipment at 0.98 million tonnes, vehicles at 0.86 million tonnes, and finally the agricultural industry at 0.66 million tonnes of plastics. The remaining quantity is attributable to household goods, sports/games/leisure, medical, and miscellaneous [Con21, p. 82].

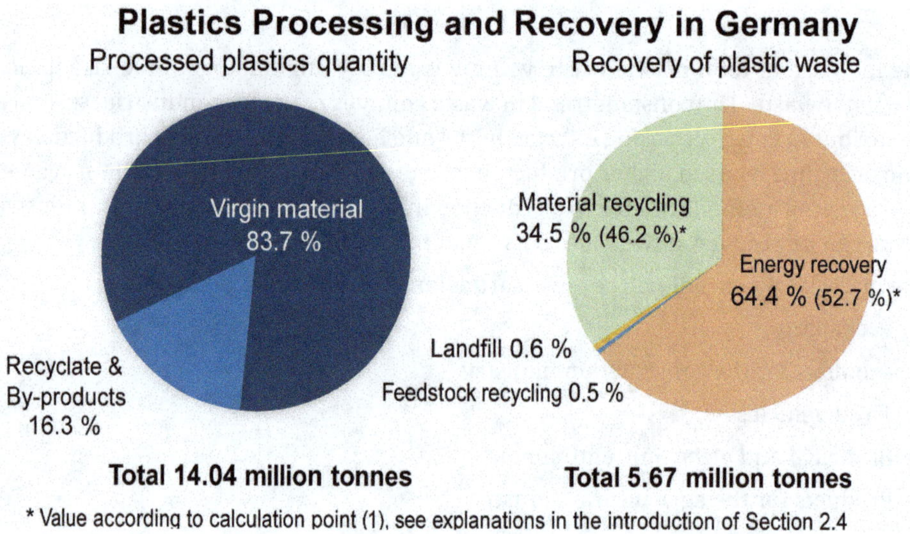

Figure 2.10 Plastics processing and recovery in Germany in 2021, figures from [Con21]

To assess the progress in the use of recyclate in plastics processing on the one hand, and in the recovery of plastic waste on the other, two diagrams are presented below for the total quantities and then for each of the key material flows already introduced. For the total quantities, the figures from 2021 from the above-mentioned study are shown in Figure 2.10.

Accordingly, in 2021 the share of recyclate and by-products in the total quantity of plastics processed amounted to 16.3%. The quantity of recyclate and by-products used increased by 17% compared to the 2019 figures, which is mainly attributable to the increased production of recyclate from post-consumer waste (approximately 25% more compared to 2019). Overall, nearly 70% of recyclates are used in the construction industry and for packaging.

The total quantity of plastic waste in 2021 amounted to 5.67 million tonnes. In addition, 0.64 million tonnes of by-products were generated, which in the previous approach of the Conversio study up to 2019 had been counted as post-industrial waste. Of the waste quantity, 34.5% was recovered by material recycling and 64.4% by energy recovery. Here too, there is a difference in the calculation basis for the respective recovery options compared to the Conversio study with figures from 2019. Previously, recycling quantities were recorded according to calculation point (1) based on the input into a recycling facility, regardless of whether material losses occurred during the shredding, sorting, and washing processes. The new calculation point (2) only takes into account the quantities fed into the final processing step for the production of recyclate. The basis for this is EU Implementing Decision 2019/665. Under the new calculation point, material losses during the processing operation are counted as additional input to energy recovery. As a result, the recycling share decreases significantly compared to the previous calculation.

In order to compare the 2021 figures with those from 2019, the diagrams also indicate the values according to the previous calculation point (1) in brackets. However, for the total waste quantity, the stated share of 46.2% cannot be directly compared with the corresponding value from 2019 (46.6%), because the by-products were still counted as waste in 2019. If the quantity of by-products is notionally added to material recycling, the share of material recycling in 2021 amounts to approximately 52%. Overall, approximately 12% more input material was directed to the material recycling route compared to 2019.

It is noteworthy that the quantities in feedstock recycling continue to decline, with only 26,000 tonnes being recovered in this way in 2021. Landfilling is also declining and, at 36,000 tonnes, accounts for a share of 0.5% of the total plastic waste quantity. Over 80% of the post-consumer waste quantity that is recycled as material originates from the packaging material flow, followed by the construction industry (approximately 7%) and agriculture (approximately 5%) [Con21, p. 93].

2.4.1 Material Flow: Packaging

The largest material flow in terms of quantity, both in processing and in the amount of waste generated, is the packaging material flow. Figure 2.11 shows the quantity of plastics processed from virgin material/recyclate and the recovery of plastic waste from the packaging sector. A distinction is made between the various recovery options.

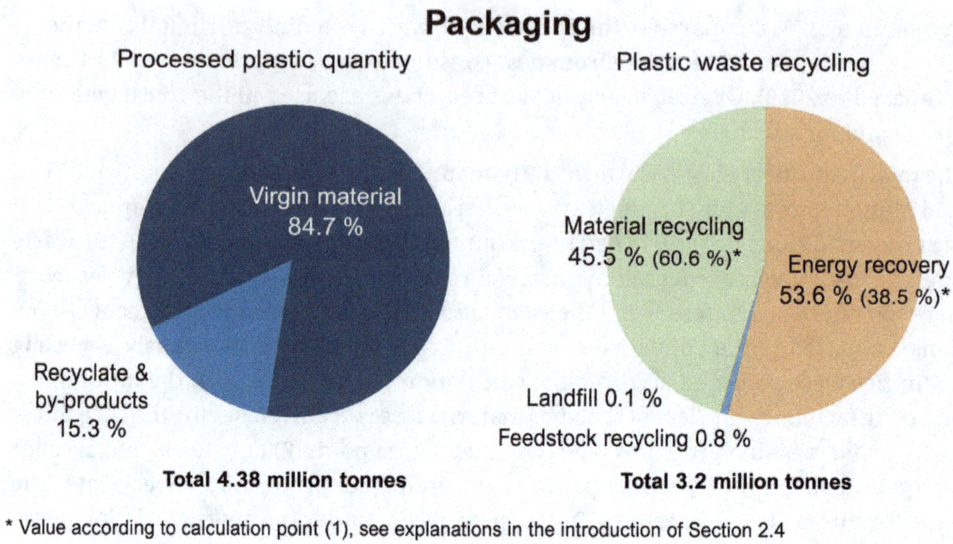

Figure 2.11 Quantity of plastics processed and recovery of plastic waste; packaging material flow in 2021, figures from [Con21]

The share of recyclate and by-products in the quantity of plastics processed in the packaging sector amounts to 15.3%, which represents a significant increase compared to 2019. While the absolute quantity of recyclate in 2019 was still 474,000 tonnes, it stood at 670,000 tonnes in 2021. This is the second largest quantity, behind that used in the construction industry. Nevertheless, the share of recyclate is below average compared to the figures for total plastics processing (see Figure 2.10). At approximately 40%, films account for the largest share of packaging products. Film production takes place on large-scale plants with high take-off speeds, which requires a high degree of process control. Processors therefore frequently rely on virgin material in order not to jeopardize production through material-related disruptions. Furthermore, to date only the PET fraction from the deposit collection system, after processing in approved production facilities, has been approved for direct food contact. As a result, packaging in this segment, apart from PET bottles and inner layers in multilayer applications, is supplied exclusively with virgin material. With 160,000 tonnes of recyclate, the largest share is used for PET beverage bottles, followed by films for the transport and shipping sector, technical packaging, transport boxes, and pallets.

In the recovery of packaging waste, a significant increase compared to 2019 can be observed in the share directed to material recycling (calculation point 1). While this stood at 48.8% in 2019, it had already reached 60.6% by 2021. The quantity actually passing through the final processing step accounts for a share of 45.5% of the total waste quantity in this material flow in 2021. Feedstock recycling accounts for a small share of 0.8%, although it should be noted that packaging waste is the only material flow that feeds feedstock recycling. The various recovery routes for household-related and commercial packaging are shown in Figure 2.12.

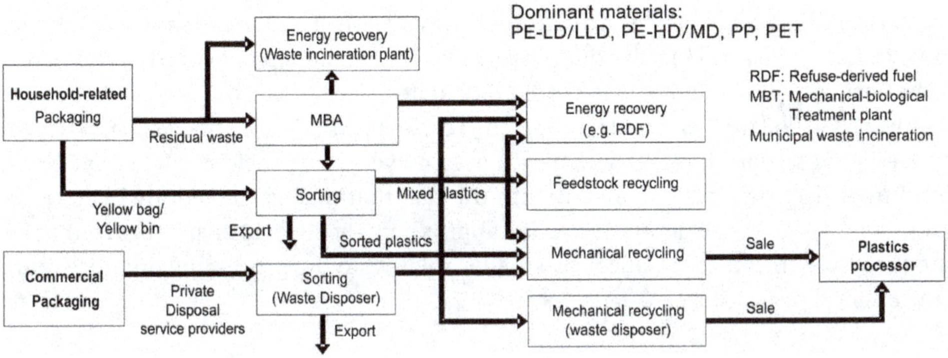

Figure 2.12 Recovery routes for household-related and commercial packaging according to [Bri19, p. 236]

Commercial packaging is largely easy to sort and only slightly contaminated. Material recycling therefore works particularly well in this case, carried out either by the waste disposal company itself or by third parties. The resulting secondary raw material is then marketed to plastics processors. In addition to domestic recovery, a portion of this waste is also exported. In comparison, household-related packaging is recovered by material recycling, energy recovery, or feedstock recycling. The variety of recovery routes for household-related packaging is attributable to the heterogeneous waste streams. In addition to the yellow bin or yellow bag, this packaging also enters waste recovery via residual waste. Furthermore, packaging waste frequently consists of small-sized, mixed, and contaminated plastics, which are sometimes also inadequately separated by consumers. Material recycling requires elaborate sorting, cleaning, and processing, which often makes this recovery route uneconomical, and energy recovery comes into play.

2.4.2 Material Flow: Products from the Construction Industry

The construction industry, with a processing quantity of 3.69 million tonnes, accounts for another large share of plastics consumption in Germany. Pipes, profiles (e.g., window profiles), and insulation materials are the main applications. The recyclate content of all construction components is nearly 25%, which is the highest compared to the other material flows after the agricultural industry. Measured against the total quantity of recyclate and reused by-products processed in Germany, the construction industry uses the most recyclate (910,000 tonnes). The largest share is used for road construction products such as posts, traffic cones, or delineator bases, followed by window and door profiles and pipes [Con21, p. 75].

As can be seen from Figure 2.13, the quantity of plastic waste generated in 2019, at 550,000 tonnes, is well below the quantity of plastics processed in the same year. If the waste quantity is related to the quantity of plastics consumed or placed on the market in Germany in the same year within this material flow, this difference decreases slightly by 330,000 tonnes [Con21, p. 81]. The reason for this large difference is that the service life of the products, at sometimes over 40 years, is considerably longer compared to packaging. This, combined with the fact that plastics have only been used extensively in the construction sector for approximately 30 to 40 years, means that to date the waste stream is significantly smaller relative to the processing quantity than is the case for packaging.

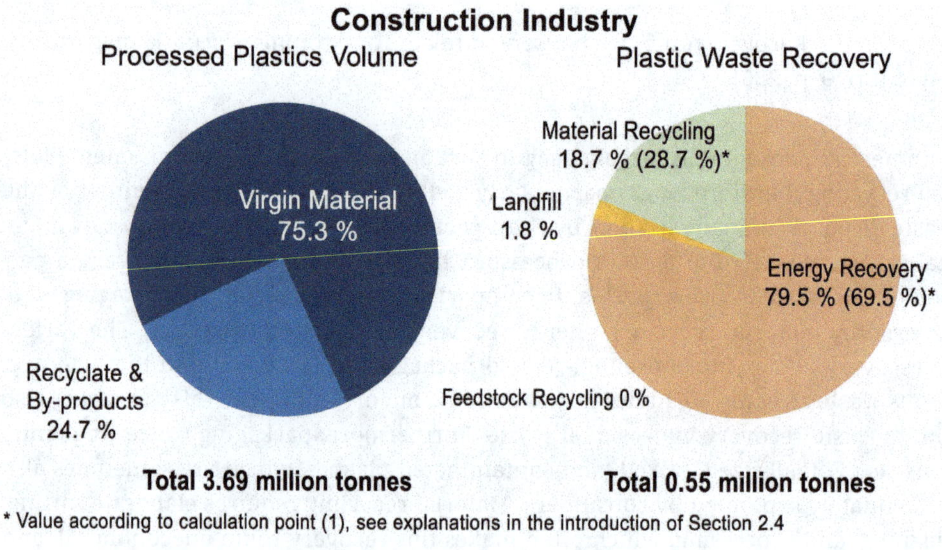

Figure 2.13 Quantity of plastics processed and recovery of plastic waste; material flow of products from the construction industry in 2021, figures from [Con21]

The treatment of plastic waste from the construction sector is dominated by energy recovery. Adhesions of mortar and adhesive residues preclude material recycling in many cases. A further challenge lies in the collection and sorting of waste, which mainly originates from building demolitions. Although pre-sorting takes place on the construction site, predominantly windows and doors are sorted out here. Due to the firm bonds between mineral and plastic materials, separation into single-grade fractions is not possible.

Figure 2.14 shows the collection and recovery routes for plastic waste from building demolitions. Single-grade plastic collections, e.g., PVC windows, can be processed and used, for example, as new PVC products or as recyclate for road construction products. Non-single-grade plastic collections and mixed collections with other construction waste must be pre-sorted. Depending on the degree of purity, the plastic is either disposed of in a landfill or recovered by material recycling or energy recovery.

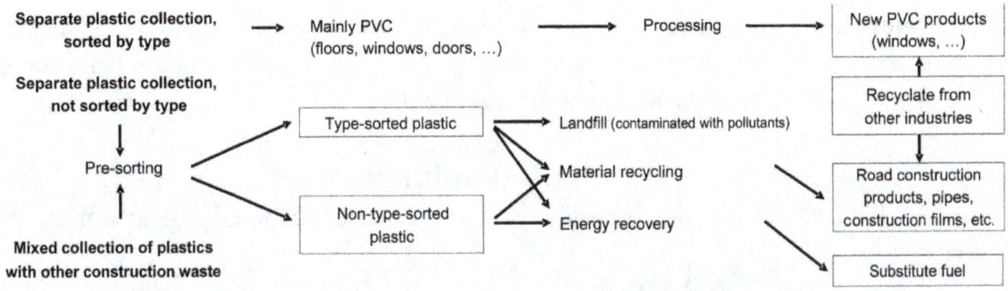

Figure 2.14 Collection and recovery routes for waste from the construction industry [Vin15]

After material recycling, the plastic can be used for road construction products, pipes, or construction films. If the waste is recovered energetically, it serves as a substitute fuel.

Compared to the other material flows, the quantity of waste that is landfilled is the largest. This is mainly due to waste contaminated with hazardous substances, for example from insulation materials, which may not be recovered by other means.

2.4.3 Material Flow: Tired Vehicles

Another important material flow is that of tired vehicles, which primarily refers to passenger cars and trucks. With a processed plastics quantity of 1.23 million tonnes, the automotive industry represents the third largest material flow for plastic products in Germany. Characteristic of this sector is the large number of different plastic types, which are frequently provided with fillers and reinforcing materials and are in some cases characterized by a surface finish (painting or metallization). As can be seen from Figure 2.15, only 5.7% recyclate is used in the automotive sector, for example in the manufacture of bumpers, interior grab handles, or carrier components for decorative elements [Con21, p. 78]. Recyclate generally cannot be used for applications with high optical requirements. The same applies to components with tight specifications regarding mechanical properties.

The quantity of plastic waste in this material flow, at 200,000 tonnes, is significantly below the quantity of plastics placed on the market in Germany in the same year within this material flow, which amounts to approximately 850,000 tonnes [Con21, p. 81]. As explained for the construction industry, this is partly due to the comparatively long service life of vehicles combined with the fact that significantly more plastics have been incorporated into vehicles over the last 20 to 30 years. However, the decisive reason for the low waste quantity is that only approximately 19% of vehicles deregistered annually in Germany are recovered domestically [Con21, p. 86].

The majority of vehicles are exported and continue to be used abroad. These vehicles are therefore not available to the German waste stream. In the recovery of plastic waste, energy recovery dominates with nearly 79% of the total waste quantity. Only 19.5% of the waste is recovered by material recycling.

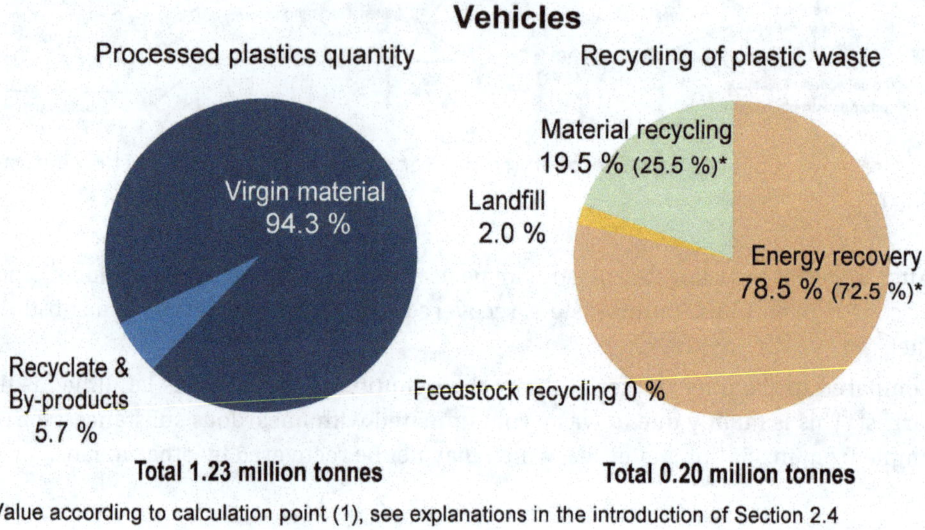

Figure 2.15 Quantity of plastics processed and recovery of plastic waste; material flow of tired vehicles in 2021, figures from [Con21]

The high share of energy recovery is attributable, among other things, to the material mix in vehicles. The focus is initially on the treatment of metal scrap, which stands out in terms of weight. Figure 2.16 The figure shows the recovery routes for vehicles. Before end-of-life vehicles are compressed and subsequently shredded as part of the recovery process, operating fluids are removed and the battery, tires, and large plastic components are dismantled.

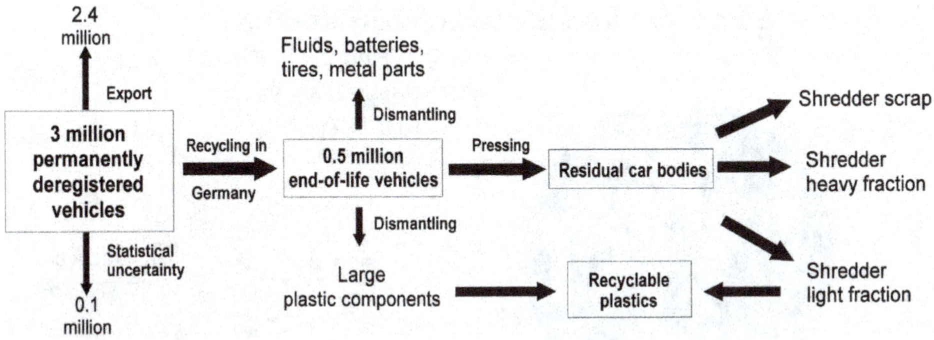

Figure 2.16 Recovery routes for end-of-life vehicles, data from [NN11-ol, NN12-ol]

The fractions after shredding are divided into shredder scrap, the shredder heavy fraction, and the shredder light fraction. From the latter, theoretically recoverable plastics can be obtained.

Recovery facilities are required to dismantle large plastic components, such as bumpers or the dashboard, prior to shredding if they are unable to separate the recyclable materials from the light fraction for material recycling after shredding. After shredding, however, the material flow is present in a heterogeneous state, so that this option is only possible with considerable technical effort. Energy recovery is currently the most economical recovery route, since the quality of the plastic fraction is insufficient and further sorting is therefore not economically viable [NN13].

2.4.4 Material Flow: Electrical and Electronic Equipment

With a processed plastics quantity of 900,000 tonnes, the electrical and electronic equipment material flow accounted for a share of 6.2% of the total quantity of plastics processed in Germany in 2021. Relative to the processing quantity within the material flow, the recyclate content is only 4.5%; relative to the total quantity of recyclate used in Germany, the share in this material flow is approximately 2% [Con21, p. 60]. Compared to the quantity of recyclate including by-products used in the electronics sector in 2019, an increase of 29% is recorded in 2021.

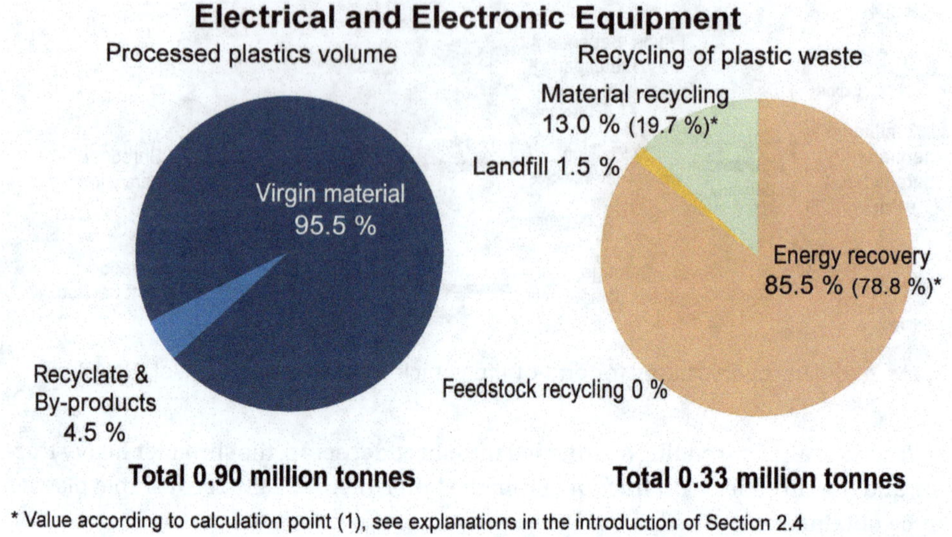

Figure 2.17 Processed plastic quantity and recycling of plastic waste; material flow of electrical and electronic equipment in 2021, figures from [Con21]

In general, many different plastics are used within this material stream, among which LDPE, PP, ABS and PA represent the most important in terms of quantity. The variety of materials results from the widely differing areas of application of plastics in the four most important areas of application in the electrical industry according to the Conversio study: cables/installation technology, white goods, IT/telecommunications and brown goods [Con21, p. 74].

The plastics used frequently contain flame retardants in order to meet applicable safety standards. In combination with the substances commonly found in waste electrical and electronic equipment, such as coolants, mercury or lead, e-waste has a high pollutant potential. For this reason, the placing on the market, the take-back and the disposal are subject to numerous regulations in accordance with the Electrical and Electronic Equipment Act (ElektroG) of 20.10.2015. For various reasons, waste electrical and electronic equipment nevertheless takes other collection and recovery routes, in which material recovery plays a subordinate role. As can be seen from Figure 2.17, only 13 % of the plastic waste installed in waste electrical and electronic equipment is recycled as a material. Waste contaminated with pollutants ends up in landfill and the majority of the waste is recovered for energy [Con21, p. 92].

Figure 2.18 shows the principal recycling paths for waste electrical and electronic equipment in Germany. Small appliances in particular are in many cases improperly disposed of via household waste and then end up in energy recovery. In addition, there are collection points that must be offered by the municipalities. Alternatively, take-back is carried out by manufacturers or retailers. Devices that are still functional can be resold after an inspection, which frequently takes place abroad. However, as determined by the Federal Environment Agency, devices that are no longer functional also end up illegally abroad through false declaration as second-hand goods [NN12-ol].

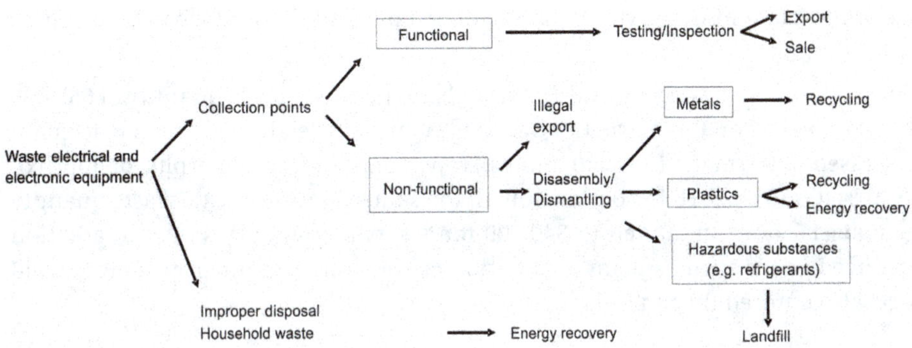

Figure 2.18 Recycling paths for waste electrical and electronic equipment, own illustration based on [MG16], [NN13]

In the case of proper disposal, the step of hazardous substance removal and valuable material separation takes place in a certified initial treatment facility in accordance with ElektroG, § 20, Para. 2. In this way, plastics are also separated out and either materially recycled or sent for energy recovery. In the area of brown goods, black plastics present a problem during sorting, because they cannot be detected, or can only be inadequately detected, by near-infrared spectroscopy. A further challenge exists in the recycling of flame-retardant plastics, because effective flame retardancy can no longer be guaranteed due to unknown ageing processes, or undesirable side effects may occur in the second life phase. For this reason, flame-retardant plastics are frequently sent for energy recovery.

2.4.5 Material Flow Products for the Agricultural Industry

The last area to be mentioned within the five largest material flows concerns the agricultural industry, which includes horticulture and landscaping as well as agriculture and forestry. With a recyclate share of 41.5% of the processed plastic quantity of approximately 600,000 tonnes, the agricultural industry used the most recyclate in the manufacture of products relative to the production volume in 2021 compared to the other material flows shown (see Figure 2.19). This means that the recyclate share has increased by 5 percentage points compared to 2019. Also in terms of the total quantity of recyclate processed in Germany, the agricultural industry ranks third with 11%, after the packaging and construction industries. Taken together, 80% of the total recyclate quantity is processed in these three material flows [Con21, p. 60].

In the agricultural sector, plastics are used, among other things, for films (mulch film, silage film, etc.), nets as protective systems in fruit and vegetable cultivation, plant pots, products for irrigation technology, greenhouse roofing, as well as in all kinds of applications in the area of stable and pasture farming. The dominant products for the use of recyclate in the agricultural sector are plant and flower pots, followed by agricultural films.

In 2021, products in the agricultural sector with a processed plastic quantity of 655,000 tonnes were placed on the market in Germany, which, in relation to the plastic quantity processed in Germany (600,000 tonnes), represents an import surplus of approximately 10% [Con21, p. 81]. In comparison, in the same year the plastic waste quantity in this material flow amounted to 340,000 tonnes, which indicates that, in addition to short-lived goods such as plant pots or non-reusable films, numerous longer-lived products also entered the market.

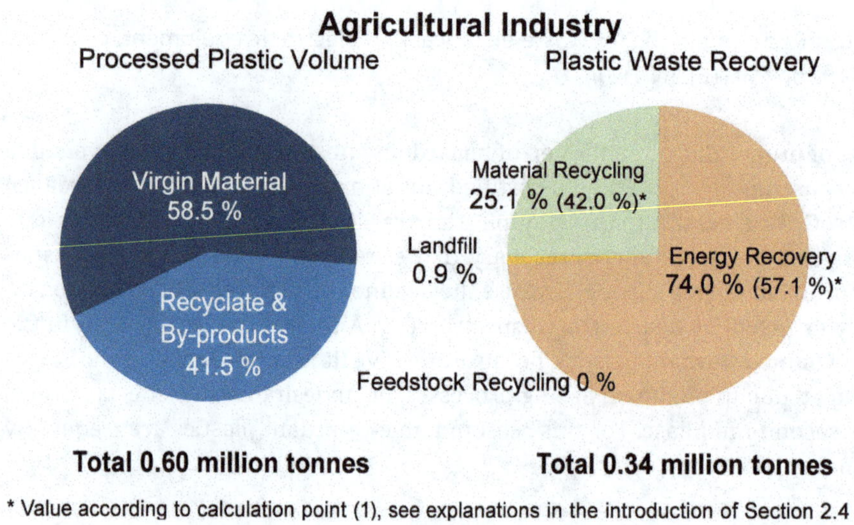

Figure 2.19 Processed plastic quantity and recycling of plastic waste; material flow products for the agricultural industry in 2021, figures from [Con21]

As with plastic packaging waste, plastic waste within this material stream also shows that a large proportion is fed into material recycling (42% after calculation point (1)). This is due, among other things, to the activities of the so-called ERDE initiative, which is organised under the umbrella of the association IK Industrievereinigung Kunststoffe e. V. and was developed with the system operator RIGK. ERDE is a take-back and recovery system for so-called agricultural plastics, essentially silage and stretch films. The system is financed by the manufacturers and distributors of the films who are members of the initiative. In 2019, ERDE submitted a voluntary commitment to the Federal Ministry for the Environment, stating that by 2022, 65% of all stretch and silage films placed on the German market would be collected and recycled [NN11-ol]. As in Section 2.3.1 as clarified under the section on self-commitment, an important building block for the effectiveness of a self-commitment is monitoring by independent experts, which was also employed in this self-commitment.

The principle of the ERDE initiative shows Figure 2.20 The farmers bring the pre-cleaned films to collection points, where the films are sorted. These are transported via disposal companies to recycling companies, which produce recyclate for the renewed production of film. In 2022, just under 38,500 tonnes of film waste were collected via the ERDE collection points and fed into material recycling, which corresponds to a share of 68 percent of the silage and stretch films placed on the market in Germany [NN14-ol; NN15-ol].

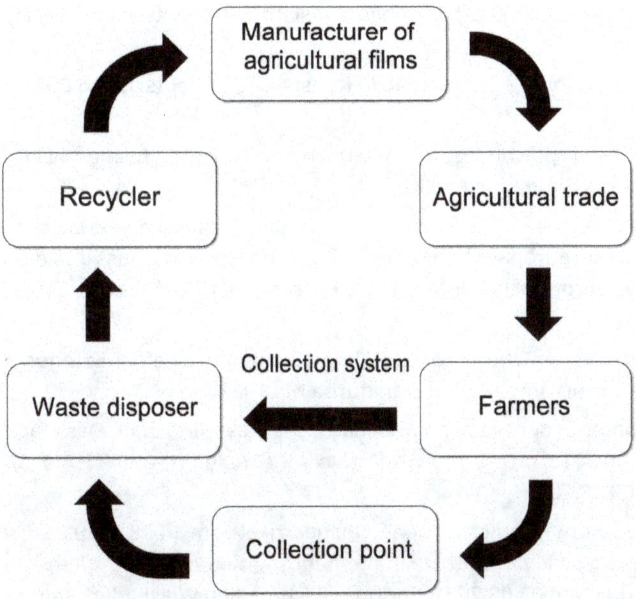

Figure 2.20 Recycling pathways for agricultural films according to the ERDE initiative (own illustration)

However, the significance of the new calculation point (2) becomes clear for this material flow. In 2021, 142,000 tonnes of waste were fed into material processing, but only 85,000 tonnes reached the final processing step for recyclate production. This means that the material loss in the processing plants for material recovery of waste from the agricultural industry is approximately 40%. This is due to the high degree of contamination of the waste from this material flow.

References

[Ado08] Adolf, J.: „Marktwirtschaftliche Instrumente – Königsweg der Klimapolitik?". Wirtschaftsdienst, ISSN 0043-6275, Springer, Heidelberg, Vol. 88, Issue: 5, pp. 326–333, http://dx.doi.org/10.1007/s10273-008-0799-1, 2008

[Bri19] Briese, D.: „Markt für Kunststoffrecycling in Deutschland bis 2025", in: Thiel, S.; Holm, O.; Thomé-Kozmiensky, E.; Goldmann, D.; Friedrich, B. (Hrsg.): Recycling und Rohstoffe – Band 12. Thomé-Kozmiensky Verlag, Neuruppin 2019

[Can01] Canier, D.: „Selbstverpflichtung der Wirtschaft und Einhaltung der Mehrwegquote". In: Wirtschaftsdienst, Vol. 81 (7), pp. 389–394, 2001

[CdA20] DeArmitt, C.: „The Plastics Paradox – Fakten für eine bessere Welt". Phantom Plastics LLC, 2020

[Con19] Conversio Market & Strategy GmbH: „Stoffstrombild Kunststoffe in Deutschland 2019", August 2020

[Con21] Conversio Market & Strategy GmbH: „Stoffstrombild Kunststoffe in Deutschland 2021", November 2022

[Eic03] Eickhof, N.: „Freiwillige Selbstverpflichtungen aus wirtschaftswissenschaftlicher Sicht". Diskussionsbeitrag Nr. 61, Universität Potsdam, 2003

[EU19] Europäische Kommission: „Mitteilung der Kommission an das europäische Parlament, den europäischen Rat, den Rat, den europäischen Wirtschafts- und Sozialausschuss und den Ausschuss der Regionen – Der europäische Grüne Deal", Brüssel, 11.12.2019, COM (2019) 640 final

[Hab62] Habermas, J.: „Strukturwandel der Öffentlichkeit – Untersuchungen zu einer Kategorie der bürgerlichen Gesellschaft". Suhrkamp Verlag, Frankfurt a.M., 1962

[Kai13] Kaiser, J.: „Die Rolle des Staates in der Grünen Wirtschaft". Die Volkswirtschaft, Das Magazin für Wirtschaftspolitik, 01.06.2013, https://dievolkswirtschaft.ch/content/uploads/2013/06/kaiser.pdf, abgerufen am 09.08.2022

[Lei19] Leitherer, J.: „Marken müssen um Händler buhlen". Onlineartikel vom 14.08.2019, https://www.springerprofessional.de/handel/vertriebsmanagement/marken-muessen-um-haendler-buhlen/17049772, abgerufen am 03.09.2021, Springer Fachmedien Wiesbaden GmbH, 2019

[MG16] Martens, H.; Goldmann, G.: „Recyclingtechnik", Springer Fachmedien, Wiesbaden, 2016

[MHL20] Mederake, L.; Hinzmann, M.; Langsdorf, S.: „Hintergrundpapier: Plastikpolitik in Deutschland und der EU – Aktuelle Gesetze und Initiativen". Ecologic Institut, Berlin, 2020

[Mic96] Michaelis, M.: „Ökonomische Instrumente in der Umweltpolitik", ISBN 13:978-3-7908-0916-9, Physica-Verlag, Heidelberg, 1996

[Nei07] Neidhardt, F.: „Öffentlichkeit und öffentliche Meinung". In: Fortschritte der politischen Kommunikationsforschung, Hrsg.: Krause, B.; Fretwurst, B.; Vogelsang, J., VS Verlag für Sozialwissenschaften, Wiesbaden, 2007

[NN01-ol] Bundesministerium für Umwelt, Naturschutz und nukleare Sicherheit, Infografik: Leitbild der Kreislaufwirtschaft, https://www.bmu.de/media/leitbild-der-kreislaufwirtschaft, abgerufen am 01.10.2021

[NN02-ol] Bundesregierung: „Dünne Plastiktüten künftig verboten". https://www.bundesregierung.de/breg-de/suche/dunne-plastiktueten-verboten-1688818, abgerufen am 01.09.2021

[NN03-ol] Cosmetics Europe: "Leading voluntary actions".https://cosmeticseurope.eu/how-we-take-action/leading-voluntary-actions/, abgerufen am 01.09.2021

[NN05-ol] FocusOnline-Artikel: „Händler gegen Hersteller: Preiskampf im Einzelhandel: Im November drohen leere Regale im Supermarkt" vom 24.09.2020, https://www.focus.de/finanzen/news/aldi-und-lidl-edeka-rewe-wieso-im-november-leere-regale-im-supermarkt-realitaet-sein-koennten_id_12315053.html, abgerufen am 03.09.2021

[NN06-ol] Nestlé Deutschland AG: „Weniger Plastik ist nicht genug". https://www.nestle.de/plastik, abgerufen am 03.09.2021

[NN07] Hauptverband des Deutschen Einzelhandels: „Fakten zur Angebots- und Nachfragemacht von Handel und Industrie". Positionspapier, Berlin, https://einzelhandel.de/index.php?option=com_attachments&task=download&id=832, abgerufen am 09.08.2022

[NN08-ol] https://www.aldi-nord.de/unternehmen/verantwortung/ressourcenschonung/verpackungen/plastic_bank.html, abgerufen am 03.09.2021

[NN09] Kunststoff Information, Vol. 51, (2563), pp. 6, 6. April 2021, S. 6

[NN10-ol] Presseinformation PreZero: PreZero beteiligt sich mehrheitlich an der KBG-Gruppe. Hamburg, 21.09.2021, https://prezero-international.com/wp-content/uploads/2021/09/Pressemitteilung-PreZero-beteiligt-sich-mehrheit-lich-an-der-KBG-Gruppe-.pdf, abgerufen am 28.09.2021

[NN11-ol] Umweltbundesamt: „Altfahrzeuge in Deutschland", https://www.umwelt-bundesamt.de/themen/abfall-ressourcen/produkt-verantwortung-in-der-abfallwirtschaft/altfahrzeuge#altfahrzeuge-in-deutschland, abgerufen am 19.10.2021

[NN12-ol] Umweltbundesamt: „Jahresbericht über die Altfahrzeug-Verwertungsquoten in Deutschland im Jahr 2019". https://www.bmu.de/fileadmin/Daten_BMU/Download_PDF/Verkehr/jahresbericht_altfahrzeug_verwertungsquoten_2019_bf.pdf, abgerufen am 23.12.2021

[NN13] Umweltbundesamt: „Export von Elektroaltgeräten". Hintergrundpapier, Juli 2010

[NN14-ol] ERDE-Initiative: „Freiwillige Selbstverpflichtung zur Rücknahme und Verwertung gebrauchter Agrarfolien". https://www.erde-recycling.de/fileadmin/user_upload/ERDE_recycling.de/Dateien/PDF-Dokumente/Finale_freiwillige_selbstverpflichtung_26.6.2019_mit_anlagen.pdf, abgerufen am 27.01.2021

[NN15-ol] ERDE-Initiative: „ERDE-Recycling erfüllt freiwillige Selbstverpflichtung und sammelt über 68 % der Silo- und Stretchfolien". https://www.erde-recycling.de/aktuelles/erde-recycling-erfuellt-freiwillige-selbstverpflichtung-und-sammelt-ueber-68-der-silo-und-stretchfolien/, abgerufen am 14.07.2023

[Ree06] Reese-Schäfer, W.: „Politische Theorie der Gegenwart in achtzehn Modellen". Oldenbourg Wissenschaftsverlag, München, 2006

[RRR+08] Rennings, K.; Rammer, C.; Oberndorfer, U.; Jacob, C.; Boie, G.; Brucksch, S.; Eisgruber, J.; Haum, R.; Mußler, P.; Schossig, C.; Vagt, H.: „Instrumente zur Förderung von Umweltinnovationen". Bericht zum Forschungsprojekt, Umweltbundesamt, ISSN: 1865-0538, Berlin, 2008

[SR03] Sprenger, U.; Rave, T.: „Berücksichtigung von Umweltgesichtspunkten bei Subventionen – Bestandsaufnahme und Reformansätze". Texte 30/03 des Umweltbundesamtes, Berlin, 2003

[Söl02] Söllner, F.: „Die Selbstverpflichtung als umweltpolitisches Instrument". In: Wirtschaftsdienst, Vol. 82 (8), pp. 478–485, DOI: 10.1007/s10273-021-2968-4, 2002

[Tis94] Tischler, K.: „Umweltökonomie". ISBN 3-486-23049-2, Oldenbourg Verlag, München 1994

[WL17] Walcher, D.; Leube, M.: „Kreislaufwirtschaft in Design und Produktmanagement – Co-Creation im Zentrum der zirkulären Wertschöpfung". Springer Gabler, Wiesbaden, 2017

3 Processing the Plastic Fraction from Waste

For waste treatment, a varying number of processing steps is required depending on the type, origin, and nature of the waste. While post-industrial waste may in the best case require neither sorting nor cleaning steps, post-consumer waste is frequently sorted and cleaned through elaborate processes. In the commercial sector, there are numerous collection and sorting routes. In the following, the collection and sorting processes along with the corresponding plant technology are first presented using the waste fraction from the yellow bag or the recyclables bin as an example. Subsequently, the further processing steps are illustrated using selected plants and application examples.

3.1 Collection and Sorting

In Germany, waste treatment is regulated by numerous laws and regulations and organized through various systems. A distinction is made between household-level collection (bring-to-door system) and collection via central locations (drop-off system). While residual waste, organic waste, waste paper, and lightweight packaging are collected at household level via corresponding containers, waste electrical equipment, waste glass, and bulky waste, for example, are frequently subject to the drop-off system. With regard to various collection systems, collection methods, and transport methods, reference is made to the standard literature on waste management, e.g. [KC17, pp. 91–133, BH13, pp. 123–181].

Separate collection by the consumer represents a first step in the sorting of waste. Thus, plastic packaging, which occurs in large quantities in households, ends up in the yellow bag or the yellow bin. Depending on the municipality, so-called recyclables bins are also used, which may also contain non-packaging items of the same material. The collection of plastic packaging is the responsibility of the dual systems, to which this responsibility has been transferred by the manufacturers, and which commission waste disposal companies for this purpose. The waste is transported via transfer stations by truck to sorting facilities for lightweight packaging, which are also commissioned through the dual systems. The sorting result is initially guided by the statutory regulations, which are contractually agreed between the operator of the sorting facility and the dual system. In addition, the technical conditions of the sorting facility are decisive for the sorting quality and the proportion of materials that can be recycled.

3.1.1 An Overview of a Light Packaging Sorting Facility

The sorting process of a light packaging sorting facility is shown schematically in Figure 3.1. In the first step, the waste is separated by so-called bag openers, which is a basic prerequisite for sorting the waste components. This is followed by sorting by size using classification methods, such as the drum screen. The waste is conveyed in the individual size classes via conveyor belts. This is followed by air classification, which removes light and loose fractions from the conveyed material. The capture of these fractions depends on the shape, size, and density of the waste pieces. The separation of ferrous metals and in particular tinplate is carried out using magnetic separators. In several partial streams, the waste pieces pass through further stations. These include the separation of non-ferrous metals (aluminum) by an eddy current separator or the separation according to the rolling ability of the waste pieces. The latter is accomplished using so-called ballistic separators. However, in order to achieve the increasing recycling quotas in the coming years, near-infrared scanning technology is probably of the greatest importance. It enables the identification and separation of individual plastic types. Modern sorting facilities are equipped with more than 20 of these NIR scanners [GV18, p. 22]. In addition to plastic types, paper, cardboard, and beverage packaging can also be identified using this technology.
To this day, automated sorting is usually followed by manual, hand sorting for individual partial streams.

At the end of the sorting process, the following fractions from the waste stream are typically available for material recovery:

- Ferrous metals and non-ferrous metals
- Sorted plastic types (PE, PP, PS, PET)
- Films
- PPK (paper, cardboard, and cartons)
- FKN (liquid cartons)

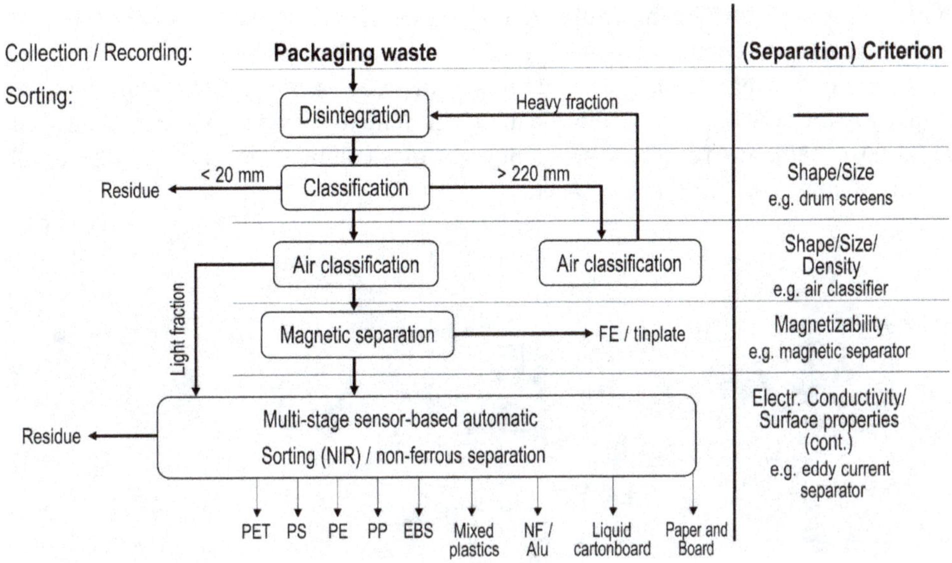

Figure 3.1 Schematic of the sorting process of a sorting facility for lightweight packaging, after [KPB+17]

The mixed plastics fraction and sorting residues remain, which are recovered for energy [TPK+19, p. 5].

The individual fractions are first stored temporarily in waste bunkers and then compacted into a stackable form using baling presses. For transfer to recycling or disposal companies, the bales of each fraction must meet certain requirements, which are set out in a product specification. This first describes the fraction in its basic composition according to material type, form, size, and permissible minor constituents. This is followed by information on the required purity of the fraction, the permissible proportions of interfering substances, and the delivery form.

3.1.2 Classification Using a Drum Screen

Classification refers to the separation of a mixture or a bulk material according to the bulk material properties of size and shape. The drum screen, also called rotary screen, serves in a light packaging sorting facility for coarse pre-sorting according to the size of the waste components. As can be seen from Figure 3.2, the drum screen consists of the drum body, whose shell is equipped with screen linings.

For use in a sorting facility for lightweight packaging, the screen is frequently designed as a perforated screen with diameters up to 200 mm. The drum has smaller screen openings in the area of the waste feed than in the rear area. This results in multi-stage screening, whereby in addition to the fine fraction in the screen underflow (SU) and the coarse fraction in the screen overflow (SO), further intermediate fractions can be discharged. The drum is mounted on running wheels, via which the drive of the rotating drum is realized. It can have a diameter of up to 3.6 m and a length of up to 18 m [KC10, p. 154]. At a rotational speed of up to 14 revolutions per minute, peripheral speeds of approx. 2.6 m/s are achieved.

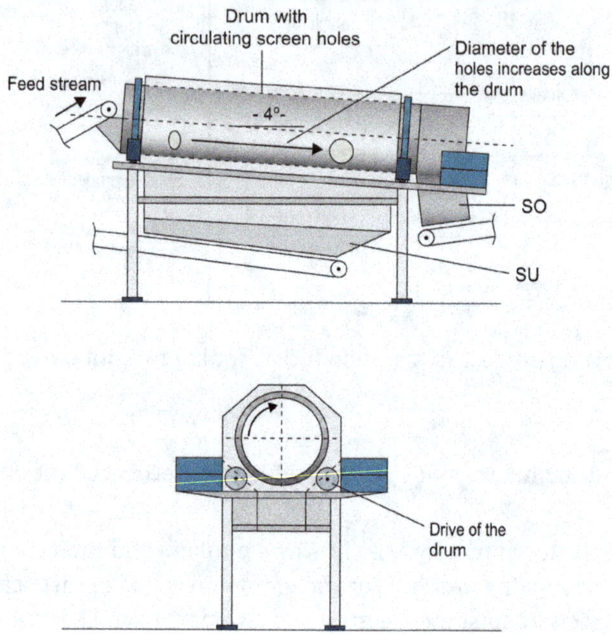

Figure 3.2 Schematic of a drum screen, after [KC10, p. 154]

Due to the inclination of the drum, the feed material moves forward in the direction of the screen overflow. The waste pieces are subjected to centrifugal force and frictional force in contact with the shell surface of the drum. This causes the waste to be carried along the cylinder wall upward until the gravitational force causes it to slide back down. The sliding along the wall can also transition into a rotating motion, causing the waste pieces to be turned over. During this up-and-down movement, the material is continuously moved forward due to the inclination. Light waste pieces, such as films, are frequently whirled up and also move in the radial direction within the drum. Through the continuous movement, the waste pieces reach the screen openings and can pass through them if the screen opening is larger than the respective waste piece. The fill level of the drum screen must not be set too high in order to reduce the accumulation of waste pieces. The residence time of the feed material in the drum can be adjusted by the degree of inclination of the drum relative to the horizontal [KC10, pp. 153 f.; BBM+95, p. 220].

Overall, the drum screen serves for the coarse classification of the waste pieces. This allows the subsequent sorting steps to be specifically designed according to the size fraction.

3.1.3 The Air Classifier

With the aid of air classification, light waste pieces such as films or foams can be separated from the remaining waste stream in light packaging sorting facilities. The separation is based on the density, size, and shape of the waste pieces. Since the waste pieces from the yellow bag are highly heterogeneous and films are also present in various size classes, the air classifiers are arranged after the drum screen (or a similarly operating pre-sorting unit). In Figure 3.3 a cross-flow air classifier with peripherals is shown schematically. The material is conveyed via a conveyor belt (1) into the intake area of the air stream generated by a fan in the classifier channel (2). Heavy material falls against the air flow onto the lower conveyor belt (3), light material is transferred via the pipeline (4) into a separator (5) and reaches a further conveyor belt (6) via a rotary valve. The air is returned via the pipeline (7) to the suction side of the fan. Approximately 30% of the air is discharged from the recirculation process via a bag filter. This allows the dust content in the air to be reduced, because on the one hand a portion of the air passes through the bag filter and on the other hand ambient air is drawn in at the inlet to the classifier channel.

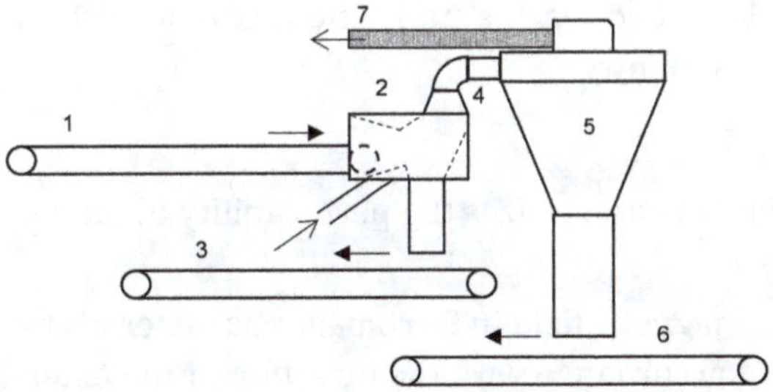

Figure 3.3 Schematic of a cross-flow air classifier

The air velocity is adjusted to the size fraction present on the conveyor belt and depending on the pre-sorted shape of the waste pieces. For air classification in light packaging sorting facilities, the projected area exposed to the airflow is frequently more important than differences in the density of the waste pieces. Films are more readily captured by the airflow due to their shape and flexibility than rigid plastic products. They can be exposed to the airflow over their full surface area and are thus discharged from the classifier channel in the direction of the airflow.

3.1.4 The Magnetic Separator

In addition to recovering recyclable plastic fractions, the task of a light packaging sorting facility is to separate the tinplate fraction from the waste. Tinplate is used, among other things together with aluminum, for canned food or beverage cans and has a high recycling rate of over 90% in Germany [NN01-ol]. Tinplate is a thin, cold-rolled steel sheet whose surface is electrolytically coated with tin. Due to its magnetic properties, tinplate can be effectively recovered from the waste stream by the magnetic separator. Magnetic separation is also necessary as an upstream unit before the eddy current separator, which would otherwise be disturbed by magnetic parts. Furthermore, it is important for compliance with the product specification of the individual sorting fractions and serves to prevent damage to downstream plant components used for separating the plastic types.

Fundamental to magnetic separation is the magnetic attractive force of a magnetic field on ferromagnetic materials. The magnetic field can be generated by a permanent magnet or induced by an electric field. The strength of the magnetic field is described by the magnetic field strength H, which in the case of an elongated current-carrying coil is defined as

$$H = n * I$$

where n the number of coil turns per coil length in the unit 1/m and I denotes the current in the unit ampere, A. The magnetic field strength thus has the unit A/m [Hin71, p. 75]. By introducing ferromagnetic materials into the interior of the coil, the magnetic effect can be amplified, which can be described via the magnetic flux density B in relation to the magnetic field strength H as follows:

$$B = \mu * H$$

Here, μ is the magnetic permeability, which characterizes the permeability of matter to the magnetic field.

A permanent or electromagnet induces a magnetic field in ferromagnetic materials by aligning so-called elementary magnets in accordance with the direction of the external magnetic field. Among the pure metals, this property applies to iron, nickel, and cobalt. The magnetic field induced in the ferromagnetic material creates attractive forces between the magnet and this material.

With regard to the application as a magnetic separator in a sorting plant, magnetic poles are thus induced in the iron-containing waste piece. The pole directed opposite to the plant magnet results in an attractive force. The magnitude of this force depends in particular on the distance of the magnet from the waste piece to be attracted, because the magnetic field strength of the magnet decreases with distance [Hin71, p. 76 ff.]. In addition, the magnetic permeability μ and the susceptibility X of the material to be attracted, which is related to it via the following relationship, also play an important role.

$X = \mu - 1$

The latter describes the individual magnetizability of a material and is significantly above zero for ferromagnetic materials. Plastics belong to the diamagnetic materials, which have a susceptibility of less than zero and may even attenuate the external magnetic field. As a result of the induction, repulsive forces occur, which are, however, small [Hin71, p. 79; DHM22, p. 423].

There are various designs of magnetic separators. In the following, only the overband magnetic separator will be discussed, because it can be well integrated into a sorting plant. It is suspended above the conveyor belts and is arranged either longitudinally or transversely to the conveying direction (see Figure 3.4). The overband magnetic separator consists of the internally located magnet and a circulating conveyor belt, which transports the waste pieces attracted by the magnet away from the waste stream to be separated onto a further conveyor belt or into a container. For the longitudinal arrangement, the magnetic separator must be positioned at one end of a conveyor belt above the head pulley. In this area, the waste stream is loosened by the onset of free fall, so that the tinplate can be attracted more effectively. The advantage over the transverse arrangement is also that the direction of movement of the parts does not change [UB97, p. 39 f.].

For the operation of the magnetic separator, pre-sorting into size classes is beneficial, because this allows the magnet to be adjusted more precisely both in terms of power and distance from the conveyor belt. Conveyor belts that are too heavily loaded can result in plastic parts covering the magnetic pieces also being attracted or clamped to the conveyor belt of the magnetic separator.

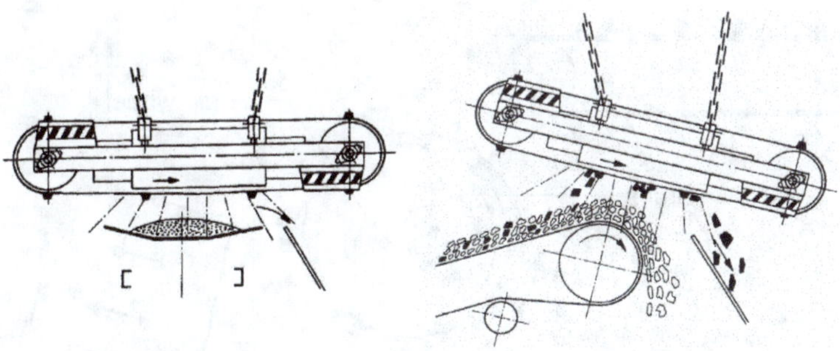

Figure 3.4 Schematic representation of a magnetic separator in transverse (left) and longitudinal arrangement (right) [UB97, p. 39/40]

3.1.5 The Eddy Current Separator

For the separation of non-magnetic metals such as aluminum, eddy current separators are used. An alternating magnetic field induces an eddy current in electrically conductive materials. This current in turn induces a further magnetic field, which, in accordance with Lenz's law, is directed opposite to the alternating magnetic field. With regard to the design in a sorting plant, the alternating magnetic field is generated either in the head pulley of a belt conveyor or in a separate drum below the head pulley of the belt conveyor by a rapidly rotating pole wheel (see Figure 3.5). The feed material passes through the effective range of the alternating field. In electrically conductive materials, such as aluminum, the eddy current and thereby the opposing magnetic field are induced around the waste piece. The resulting repulsive force leads, compared to non-electrically conductive materials such as plastics, to a different trajectory and thus to the separation of these materials. In addition to the conductivity of the material, the gravitational force acting on the waste pieces influences the trajectory. The density and conductivity are therefore the decisive material parameters for the successful discharge of electrically conductive but non-magnetic waste pieces. The higher the ratio of conductivity to density, the better the expected separation result. As already described in Section 3.1.4, a magnetic separation must precede the eddy current separator. Otherwise, the magnetic waste pieces would adhere to the drum and would thus interfere with the eddy current generation process.

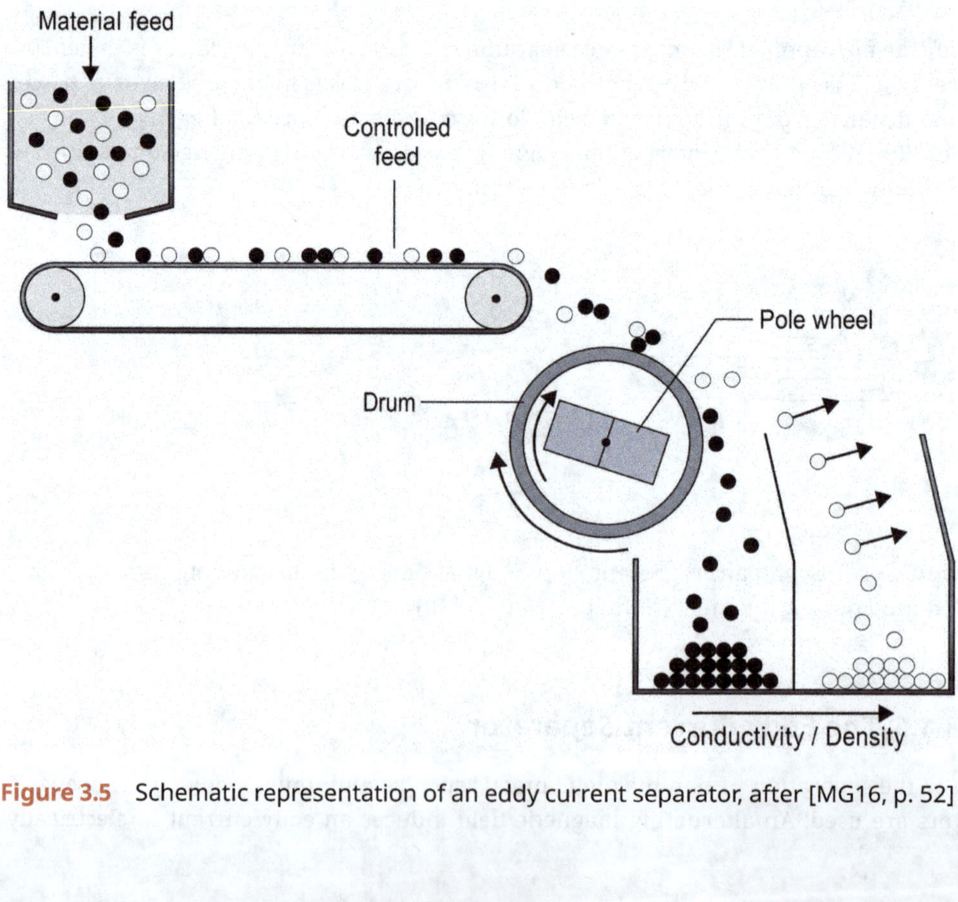

Figure 3.5 Schematic representation of an eddy current separator, after [MG16, p. 52]

3.1.6 The Ballistic Separator

The sorting process in a ballistic separator is based on the different movement behavior of waste pieces on an inclined plane in combination with impact-like vibrating movements.

The design of the ballistic separator is shown in Figure 3.6 in the side and top view. Elongated screen paddles are arranged side by side at an angle in space and form the base of the separator. They are mechanically coupled to a crankshaft or driven directly via a drive eccentric. Adjacent paddles each reach the upper apex of the circular path formed with a time offset. The feed material is placed centrally onto the paddles and is subjected to an impact load. In addition, the screen base is briefly withdrawn from the feed material during the downward movement of the paddles. It then lands further up on the paddle during the upward movement. With regard to the waste pieces in a light packaging sorting plant, the shape, elastic behavior, size, and weight are decisive for the separation of different fractions. A prerequisite for the separation effect is that the feed material is free to move.

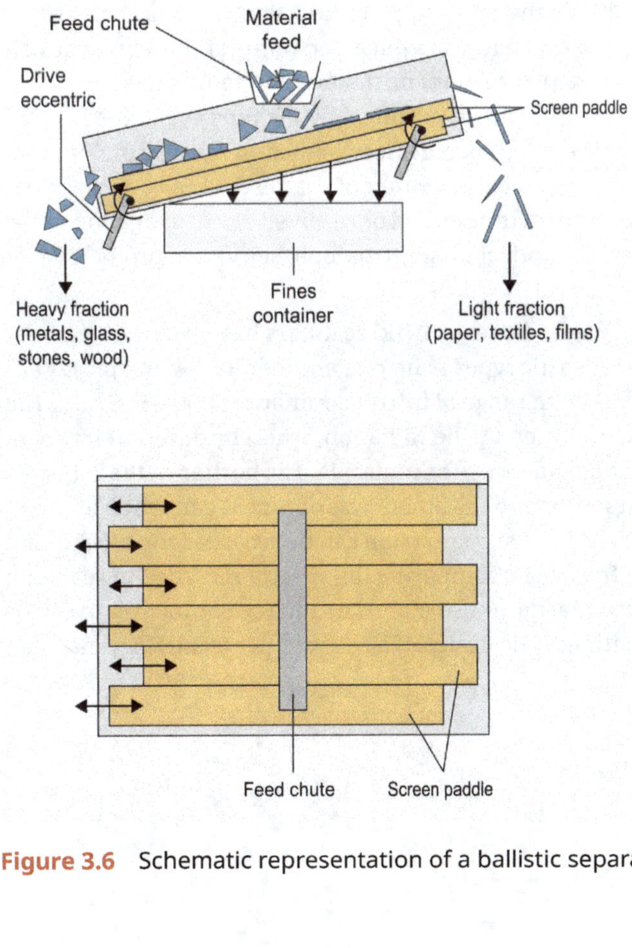

Figure 3.6 Schematic representation of a ballistic separator, after [MG16, p. 57]

Light, flat, and elastic parts migrate upward through the paddle movement and thus follow the principle of movement described above during the downward and upward movement of the paddles. Heavy, hard, and round parts experience more of an impact load from the inclined paddles upon contact and are thus moved downward. In addition, they slide or roll downward on the screen base. In this way, for example, hollow bodies such as canisters or shampoo bottles, which move downward, can be separated from small pieces of film. The perforations in the screen box serve to separate fine particles [MG16, p. 56; UB97, p. 62 f.].

The inclination of the separator, the stroke, and the frequency represent the main parameters that influence the separation result depending on the type of feed material.

3.1.7 Sensor-Based Sorting

In contrast to the processes presented so far, in which a fraction is sorted out as a collective by responding to a material property, in the process of so-called picking, the sorting criterion is checked individually on each piece and sorted out accordingly. A distinction is made between manual picking, in which personnel at the conveyor belt sort out either a contaminant (negative sorting) or the fraction to be sorted (positive sorting) individually by hand, and automated picking, in which the eye of the personnel is replaced by a sensor and the hand by a device coupled to the sensor. Material identification takes place via the specific interaction of matter with electromagnetic waves. Depending on the wavelength of the emitted or received electromagnetic wave, the interaction can be analyzed via optical or spectroscopic sensors or, in the case of X-ray radiation, via X-ray detectors [MG16, p. 58].

In light packaging sorting plants, near-infrared (NIR) scanners have been successfully used for many years to separate plastic types from one another. The waste pieces are irradiated with light in the wavelength range of infrared radiation (800–2500 nm). The polymer molecules are set into vibration by the radiation, whereby different frequencies are required for excitation depending on the molecule. The portion of the reflected radiation is detected by the NIR sensor and the intensity is output as a function of wavelength. In this way, a characteristic intensity spectrum can be recorded for each plastic and stored in the evaluation software. By comparing the spectra recorded in real time with the database, the respective plastic or also materials such as cardboard, paper, or beverage packaging can be identified. The design of a NIR scanner technology installed in a sorting plant is shown in Figure 3.7.

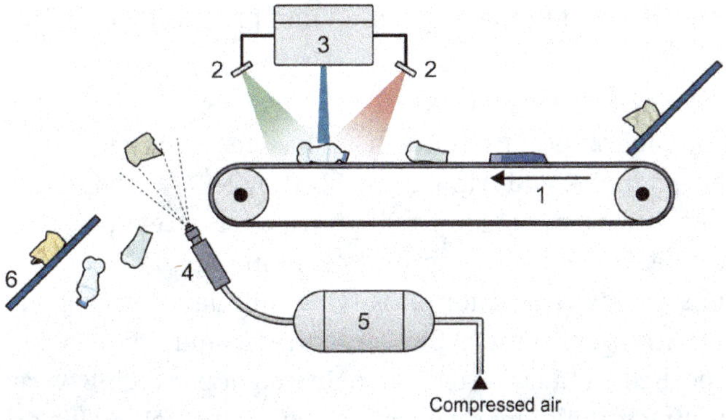

Figure 3.7 Operating principle of NIR spectroscopy installed in a sorting system, [KC10, p. 176, after Ul97]

The feed material is singulated in upstream processes and distributed across the width of the belt conveyor. It is also important to remove larger film sections that could cover other waste pieces. The belt conveyor (1) runs at a belt speed of approximately 2.5 m/s and conveys the feed material into the detection zone of the NIR scanner [BH13, p. 511; KC10, p. 175]. The waste is irradiated by two halogen or LED lamps (2), which are arranged differently depending on the design of the system. The reflected radiation is captured by the detection device (3). It is important to assign the respective spectrum to the exact position along the belt width. After identification of the material, the information on the material, the size of the waste piece, and the position of the waste piece to be sorted out — calculated in real time as a function of the belt speed — is processed. At the end of the belt conveyor, an air nozzle bar (4) is arranged, which is supplied with compressed air via a pressure tank (5). The material to be sorted out is blasted out of its trajectory behind a baffle plate (5) by the number of air nozzles assigned according to size. To improve the accuracy of detecting the shape, position, and size of the waste pieces, additional color line-scan cameras or lasers can be used [KC10, pp. 175 f., NN02]. NIR scanners cannot distinguish between plastics colored black with carbon black, because these absorb the infrared light almost completely, making it impossible to analyze any reflected radiation. Current developments are moving toward self-learning systems in which images and data on shape and color are linked together and sorting accuracy is to be improved through pattern recognition. This also includes the detection of black plastic parts when, for example, they can be specifically identified by their shape [NN02-ol]. Researchers at the Fraunhofer-Gesellschaft have developed a camera-radar system that operates in the terahertz range and with which black plastics can be directly identified [NGB+16]. The company Steinert GmbH from Cologne offers a NIR scanner with integrated hyperspectral image processing in order to directly identify black plastics as well [NN03-ol, Bee17. p. 188]. Inno-Spec from Nuremberg also offers hyperspectral cameras for use in the sorting of plastic parts.

These likewise allow a high level of detail in the acquisition and processing of information [NN04-ol].

The process-reliable detection of individual plastic types in multilayer packaging is the subject of current research. In a laboratory environment, it has been demonstrated that the differentiation of single-, double-, and triple-layer films by NIR technology is possible [CKF+20]. Whether this can also be robustly realized in a waste stream of post-consumer waste, however, must be shown by further investigations. A study by Chen et al. presents promising results in this regard for classifying multilayer packaging with different plastic types in the structure. Several film structures contaminated with oil and water and mixed with post-consumer waste from lightweight packaging were analyzed by NIR sensors. Accordingly, multilayer packaging containing polyamide or PE/PP, or containing neither PP nor PE, can be sorted out using the sensor technology employed. Very thin layers of PET could not be detected, which, according to the authors' assumption, is attributable to reflections from the glossy, thin layers [CKW+21]. The detection and sorting of fractions with multiple, material-bonded plastic types is a prerequisite for the subsequent composite separation processes required for material recycling. These processes, which currently exist only at laboratory or pilot scale, are presented in Section 4.6.

The sensor and camera systems are expensive to purchase compared to the systems presented so far, which is why the most recent technologies mentioned last are not yet widely used in today's sorting plants. However, it is to be expected that as recycling quotas are increased by legislators, more and more of the technologies already available on the market — for example for the detection of black plastics — will be employed.

3.2 Production of Recyclate

As shown in the overview of the light packaging sorting plant at the beginning of this chapter, the output of a sorting plant consists of unshredded fractions sorted by material type, shape, and size according to a defined specification, in bale form. These fractions are subjected to further processing steps depending on the nature of the fraction. Those fractions that can be recycled as material are sent to a processing or recycling company. There they typically pass through the following steps:

1. Bale opening and size reduction of the sorted waste pieces
2. Washing of the shredded plastic waste
3. Separation by plastic type during the washing process or in the dry state
4. Drying of the material stream
5. Processing or re-additivation
6. Regranulation or agglomeration

In the following, steps 1 to 4 and step 6 will be presented. Step 5 is discussed in more detail in section 4.5.

3.2.1 Size Reduction

Before the waste can be washed and finely sorted by plastic type, size reduction takes place. The bales delivered to the recycler pass through so-called bale breakers, in which a drum fitted with protruding flat iron bars typically ensures singulation of the waste pieces. A uniform discharge of the singulated waste pieces is important in order to continuously feed downstream equipment.

The main objectives pursued with the size reduction of waste are [KC10, p. 139; MG16, p. 31; Sti97, p. 225]:

- Homogenization of the waste stream by producing defined particle size distributions
- Liberation of form-fit or force-fit connections (e.g., screw caps)
- Enlargement of the surface area for the application of wetting fluids for surface treatment
- Exposure of all surfaces to be cleaned (e.g., in the case of hollow bodies)
- Enabling sorting by plastic type in downstream separation processes
- Reduction of storage space
- Production of a free-flowing intermediate product for further processing into granulate
- Material behavior

To achieve the stated objectives, different size reduction machines are used depending on the material to be processed. The background for this is the different deformation and fracture behavior of the materials, in this case plastics. As already described in section 1.3.1, plastics consist of molecular chains formed by the combination of monomers during the synthesis reaction. The cohesion of the chain molecules within the polymer chain is provided by the covalent primary valence bond and the resulting primary bonding forces. Between the polymer chains, secondary bonding forces act, which are caused by the secondary valence bonds. The following factors influence the molecular structure, the polymeric cohesion, and thus the deformation behavior of plastics:

- Type of molecule or the underlying monomers
- Molecular size (degree of polymerization, molecular mass, molecular weight distribution)
- Molecular shape (branching, rotatability about the primary valence axis)
- Molecular order (tacticity, steric order)

The resulting material behavior of plastics is complex and ranges from nearly energy-elastic to the viscoelastic behavior typical of polymers. The latter represents a superposition of energy-elastic and entropy-elastic components with viscous deformation components. The temperature of the material is a further influencing factor, because a change in temperature alters the thermal energy of the molecules, which induces molecular vibrations. The stronger the movement of the molecules, the greater their space requirement, which results in a larger distance between the molecular chains. One consequence of this is a reduction in secondary bonding forces. This gives rise to characteristic temperature ranges for amorphous and semi-crystalline plastics in which different material behavior is observed. This can be seen in Figure 3.8, for example, in the progression of the elongation at break ε_B.

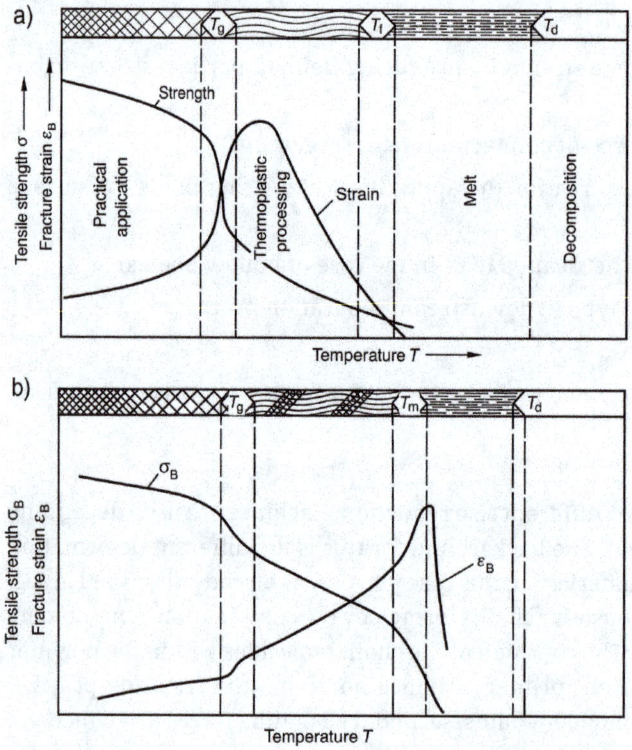

Figure 3.8 Temperature dependence of tensile strength and elongation at break for an amorphous a) and semi-crystalline b) thermoplastic [DHM22, pp. 236, 238]

Semi-crystalline thermoplastics exhibit an increase in elongation at break above the glass transition temperature, which is also reflected in the high impact strength of this group of thermoplastics. The service range of semi-crystalline thermoplastics is mostly in the range above the glass transition temperature T_g of the atactic regions, whereas in amorphous thermoplastics the glass transition coincides with the onset of melting. Below T_g thermoplastics behave in a hard-brittle manner. For crosslinking plastics, the material behavior of thermosets can likewise be described as hard-brittle, whereby the elongation at break may also increase slightly with rising temperature depending on the resin. Elastomers behave, as expected, rubber-elastically above the glass transition temperature.

From the molecular relationships explained so far, the following stress-strain curves, determined by tensile tests at room temperature, result as examples.

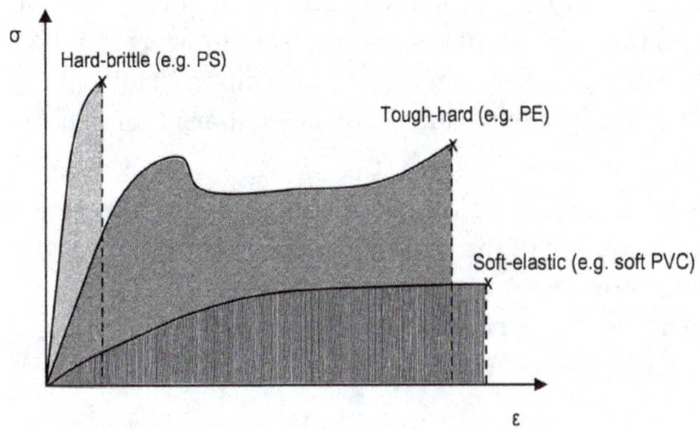

Figure 3.9 Exemplary stress-strain diagrams for plastics with different deformation behavior

For thermoplastics, due to their viscoelastic material behavior, in addition to the temperature dependence, the time dependence of the stress-strain behavior must also be taken into account. Under prolonged loading, the effect of creep causes continuous strain over time under constant external load. This occurs through time-dependent flow processes in the plastic due to the sliding of polymer chains. These flow processes are also responsible for the effect of stress relaxation, in which the stress in the plastic decreases over time at constant strain. These effects must also be taken into account even under short-term loading. If the material is given little time to relieve internal stresses through flow processes due to high loading rates, high stresses consequently occur and the material behaves in a brittle manner [DHM22, p. 241].

In summary, from the consideration of temperature and time dependence, it can be concluded that low temperatures and high loading rates lead to brittle material behavior, while high temperatures and low loading rates produce tough material behavior.

Comminution Mechanisms and Overview of Size Reduction Machines

For comminution, the required energy input per unit volume, or the comminution work up to fracture W_B is a decisive quantity. It corresponds to the area under the stress-strain curve and can thus be determined using the following equation:

$$W_B = \int_0^{\varepsilon_B} \sigma(\varepsilon)\, d\varepsilon$$

With:
W_B = comminution work
ε_B = strain at fracture

With reference to the different curves in Figure 3.9, it becomes clear that the volume-specific work for the comminution of brittle materials is small, while that for the comminution of tough-hard materials is high. From this it can already be deduced that for tough or rubber-elastic plastics, the volume that is stressed should be as small as possible in order to obtain low absolute energy amounts. This can be achieved by locally concentrated introduction of energy, for example through sharp tool cutting edges [Sti97, p. 228].

In principle, the types of stress shown in Figure 3.10 can be distinguished, which come into play primarily depending on the design of the size reduction machine. Compression, impact and percussion stresses are used for brittle materials, while these do not lead to comminution in thermoplastics. Here, cutting, shearing and tearing stresses are to be preferred.

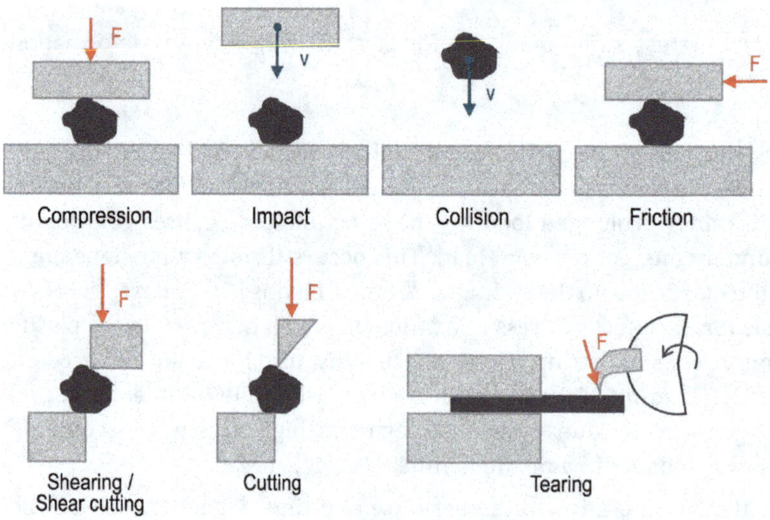

Figure 3.10 Schematic representation of types of stress, after [Sch02, p. 8; KC10, p. 139; MG16, p. 33]

It must be taken into account that, in addition to the primarily induced type of stress, superimpositions with other types of stress also occur. A multiaxial stress state of normal (tensile/compressive) and shear stresses always develops in the feed material. Depending on the ductility of the material, severe plastic deformations can occur before fracture sets in, which increases the comminution work.

The selection of machinery for the specific comminution task is made not only taking into account the fracture behavior of the feed material, but also depending on the particle size to be achieved. Various classifications are chosen in the literature. According to Kranert et al., the following particle size ranges can be defined [KC10, p. 140]:

- Coarse comminution: approx. <150 mm
- Medium comminution: approx. <50 mm
- Fine comminution: approx. <10 mm

Table 3.1 provides an overview of machines to be used for brittle and tough-elastic materials depending on the size of the particles to be achieved.

Table 3.1 Overview of Size Reduction Machines, After [KC10, p. 140]

Material property	Coarse comminution	Medium comminution	Fine comminution
Medium-hard, brittle → thermosets	Jaw crusher, impact crusher	Hammer crusher, impact crusher	Fine impact mill, ball mill
Elastic, tough-elastic →thermoplastics, elastomers	Single- or multi-shaft shredder	Single-shaft shredder, granulator, hammer mill (cryogenic)	Granulator, hammer mill (cryogenic)

Jaw crushers primarily use compression as the type of stress, while the material in impact and percussion crushers, as the name already suggests, is subjected to impact and percussion loading. The ball mill and fine impact mill apply compression to the material at low speeds and impact and percussion stress at higher speeds [Sti97, p. 289]. These machines are used for thermosets, which fracture in a brittle manner.

For thermoplastics or elastomers, machines are used that exert more of a cutting or shearing stress on the material. These are single- or multi-shaft granulators or cutting mills. In particular, multi-shaft granulators also involve tensile, bending, and torsional loads, which are collectively referred to as tearing stress [Nic96, p. 127 f.; Sch02, p. 8]. In the following, some of the machines used for the size reduction of plastic waste will be discussed in more detail.

Hammer Mill

One of the machines used for the size reduction of rather brittle plastics is the hammer mill. In addition to its main application for thermosets, it can also be used, for example, for polystyrene or rigid PVC (PVC-U) [BBM+95, p. 209]. Cold-embrittled elastomer products such as used tires can also be ground with this equipment [Jun94, p.

74]. For typical thermoplastic plastic packaging waste, however, the hammer mill is not suitable, because the energy expenditure required to reduce it would be too great [BMW97, p. 62, Jun94, p. 72].

A hammer mill, as schematically shown in Figure 3.11, consists of a horizontally mounted rotor shaft on whose circumference pivotally mounted impact or striking tools — the hammers — are attached. Additional impact elements are mounted on the housing wall, and the screen plate is located in the lower section. Hammer mills are operated at peripheral speeds of up to 100 m/s, which are high compared to cutting mills [Jun94, p. 62].

The feed material falls through the feed chute — in this case tangentially — into the working area of the hammers, where it is directly shattered and accelerated by the impact, depending on the energy absorption and cohesive strength of the particles. Fragments are flung against the impact plates, where they also break apart before returning to the working area of the hammers. In this way, the feed material is reduced in size until it can pass through the screen holes and be discharged [Sti97, p. 310].

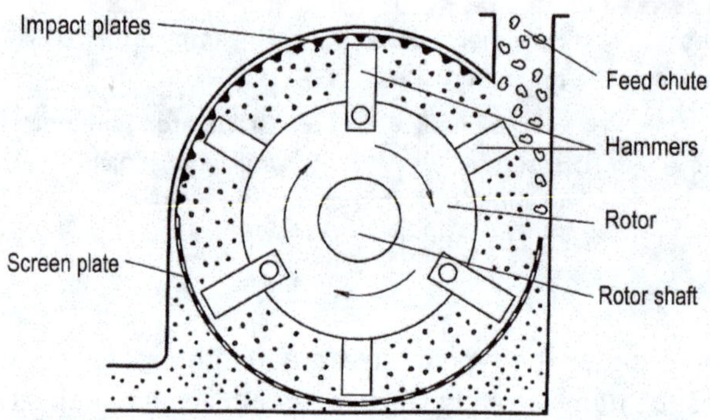

Figure 3.11 Schematic design of a hammer mill, adapted from [MBW93, p. 63]

Particularly during fine size reduction, the proportion of dust is comparatively high. In conjunction with the suction created by the rotation of the mill, extraction via filters or a cyclone is important [Jun94, p. 74].

Single- and Multi-Shaft Shredders

For the coarse size reduction of thermoplastic materials, granulators are used that consist of one or more knife-equipped shafts and are capable of reducing a wide variety of feed materials.

Depending on the size, shape, and type of feed material, the plant types shown schematically in Figure 3.12 are selected for the size reduction task.

Single-shaft shredder are used for many different waste materials, such as waste wood, electronic scrap, carpets, plastics from end-of-life vehicles, or packaging waste in bale form. They essentially consist of a horizontally mounted rotor fitted with rotor knives in various configurations depending on the material to be reduced, a hydraulically actuated pusher that presses the feed material against the rotor, a screen basket enclosing the rotor over an angle of up to 180°, and a surrounding housing. In addition to the rotor knives, so-called housing knives are arranged on the housing, maintaining a clearance of approximately 0.1 mm from the rotor knives. The peripheral speed of the rotor is between 5 and 10 m/s. The feed material enters the cutting area of the rotor knives via the hopper, with the pusher ensuring a temporally constant engagement of the knives. A cutting stress occurs between the rotor and housing knives. Once the material is smaller than the screen perforation, it is discharged through the screen basket and conveyed further by belt or screw conveyors. Throughput decreases with smaller screen perforation, as the material must pass through the knife engagement multiple times before it can pass through the screen [KC10, p. 142].

The machines are manufactured in a wide range of sizes with shaft diameters of up to 800 mm and widths of 5 m. Wear on the machines is high, for example when reducing fiber-reinforced plastics or in the presence of hard foreign matter, so that the cutting plates must either be turned or the gap width must be readjusted from time to time [Jun94, p. 143].

Twin-shaft shredder, also called rotary shears or cutting-roll pre-shredders, are used for the size reduction of large-volume plastic components. They consist of two parallel, counter-rotating rotor shafts. The cutting elements are arranged on discs that are axially offset in alternation with the parallel shaft. This allows the shafts to intermesh, which supports effective material infeed and size reduction. The shafts run at a peripheral speed of 0.3–0.8 m/s, significantly slower than single-shaft shredders. The hook-shaped cutting elements, frequently referred to in the literature as tearing teeth, draw the material into the area between the shafts, where it is reduced in size. In addition to pure shear stress in the axial gap, tensile and tearing stresses also occur. Strip-shaped sections with a size between 15 and 50 cm are frequently produced. No screening to limit particle size takes place, as is the case with single-shaft shredders, cutting mills, or hammer mills. The machines are frequently equipped with overload protection and the ability to reverse the direction of rotation. In the event that blocking contaminants are drawn in, these can be removed before they cause damage to the machine [KC10, p. 141. Jun94, p. 61; BMW97, p. 57].

Four-shaft shredder are similar in size reduction behavior to the twin-shaft shredder, are suitable for bulky feed material, and improve the infeed behavior of the machine. In contrast to the twin-shaft shredder, this type of machine is usually equipped with a screen plate, which has a large surface area due to the design. As a result, four-shaft shredders achieve high throughputs [NN05-ol, p. 4].

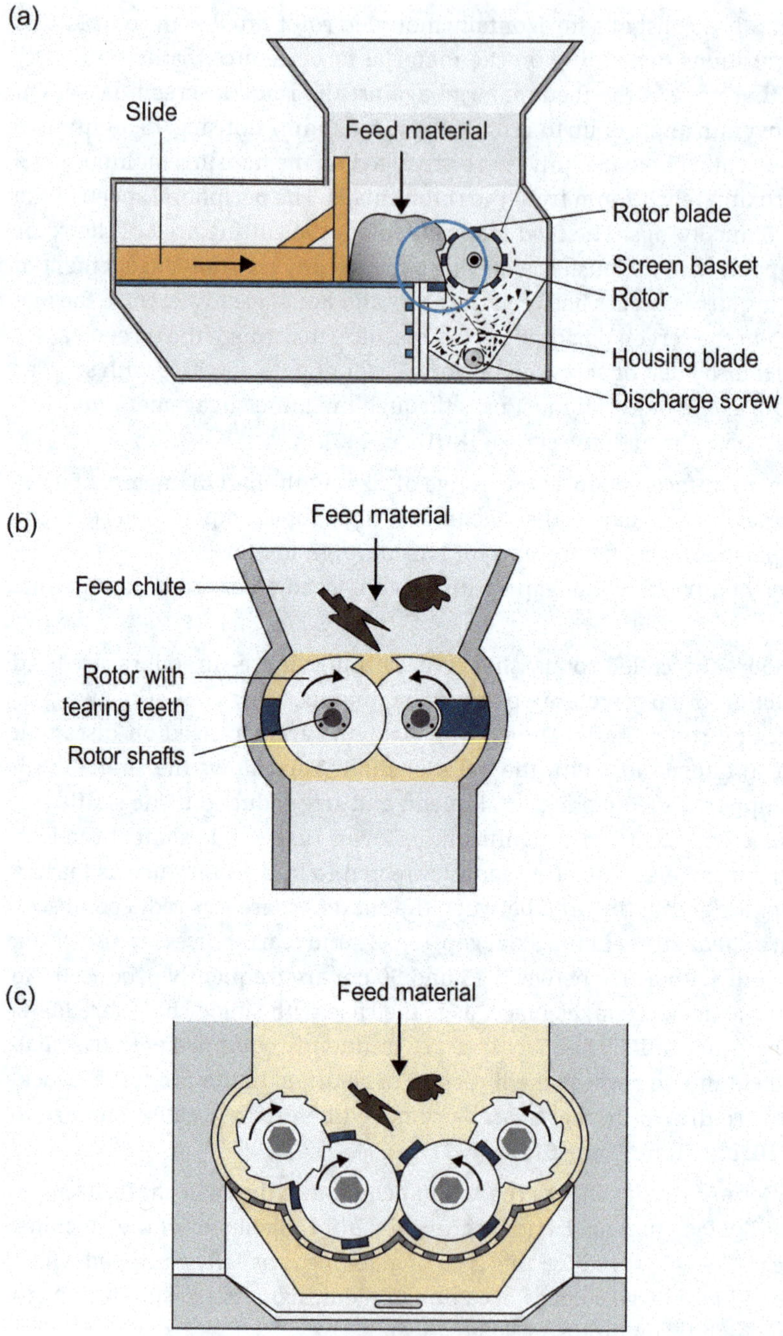

Figure 3.12 Schematic design of a) a single-shaft shredder (adapted from KC10, p. 143), b) a twin-shaft shredder (adapted from [Jun94, p. 62]), and c) a four-shaft shredder (adapted from [Sch02, p. 13])

Cutting Mill

As outlined above, for viscoelastic and rubber-elastic materials the stress should be applied to a small volume. Otherwise, the energy requirement for size reduction would be significantly higher, because a not insignificant proportion is needed for the deformation of the material. A locally acting stress is that of cutting with a defined knife edge, as predominantly occurs in cutting mills. These are operated at peripheral speeds of up to 20 m/s and are used for medium and fine size reduction. As Figure 3.13 shows, the cutting mill has a rotor on which the knife edges are bolted at distributed positions around the circumference. In some cases, these are arranged in segments and offset on multiple rotor discs. The feed material enters the cutting area of the rotor under gravity. The cutting process takes place between the fast-running rotor knife and the stator knife rigidly connected to the housing. The particles are reduced in size within the grinding chamber until they can pass through the screen plate. The design of the grinding chamber interior, the rotor with its knives, the screen plate, and the position of the feed chute are configured differently depending on the feed material. In principle, the fines content for thermoplastics is low due to the cutting stress; however, differences in the particle size distribution can be observed, for example, through the design of the screen, the arrangement of the knives, or the gap width between the rotor and stator knives [BMW97, p. 59 ff.; Jun94, p. 61; KC10, p. 144].

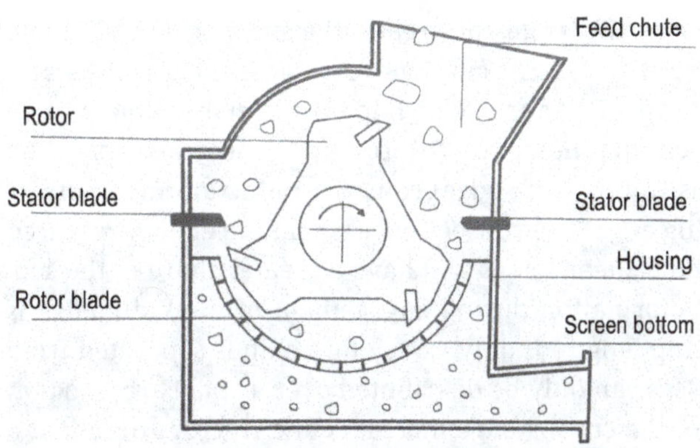

Figure 3.13 Schematic design of a horizontal cutting mill, adapted from [BMW97, p. 59]

In this case, the stator knife serves merely as a counter-bearing and is not directly involved in the cutting process. The stress situation is shown schematically in Figure 3.14. The knife edge penetrates the material in a linear fashion with a wedge angle of less than 30°, the material being clamped between the cutting edge and the stator knife. At the tip of the cutting edge, a high stress concentration develops, which leads to failure of the material in this area. In addition to the shear stress acting through the cutting edge and stator knife, a tensile stress arises, which is caused by the splitting action of the knife and the bending in the clamping situation.

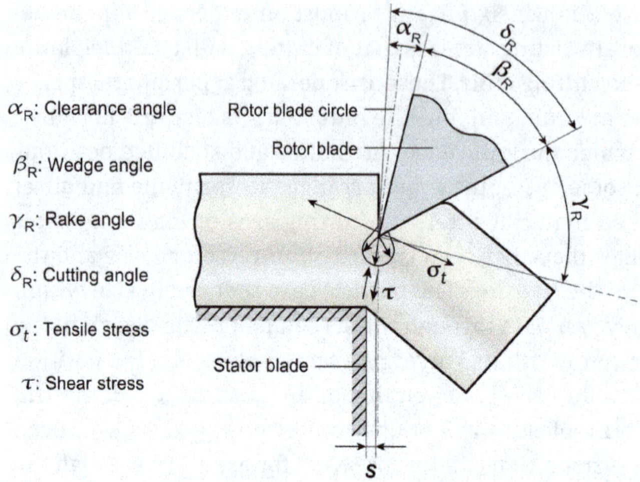

Figure 3.14 Crack initiation at the knife edge, adapted from [Wol04, p. 11]

The larger the wedge angle of the cutting edge, the greater the shear stresses become relative to the tensile stresses [Bau01, p. 33]. The cutting gap *s* also influences the stress state. The larger this is, the greater the lever arm, which in turn affects the magnitude of the tensile stress due to the bending moment. With regard to the geometry of the cutting edge, a large edge radius (blunt knife edge) causes a reduction in the stress concentration ahead of the cutting edge. The consequence is that the energy required for separation will increase and the shear forces will have a greater share in the separation process. The extent of deformation during engagement of the knife edge is additionally dependent on the chip length. If only a very thin chip is separated from the rest of the material, the stresses can only be distributed over a small area, so that the stress concentration ahead of the cutting edge increases and fracture occurs at a lower cutting force [Bau01, p. 37–39].

In principle, two types of fracture can be distinguished in their extremes, whereby plastics always exhibit a mixed form of these two [Bau01, p. 32, Wol04, p. 11]. On the one hand, there is brittle fracture, which occurs suddenly, and on the other hand, purely ductile material separation, in which the material is completely severed by cutting (see Figure 3.15).

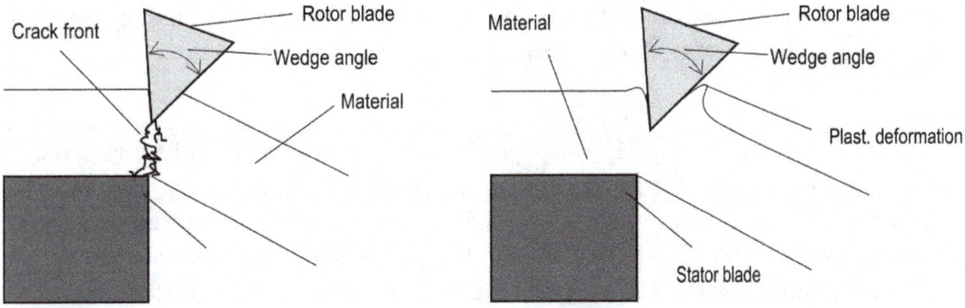

Figure 3.15 Brittle fracture and ductile material separation during size reduction by means of a knife edge

In the first case, an initial cut is made in the material, which initiates a crack. The crack front propagates suddenly and leads to brittle residual fracture in the material. In the second case, plastic deformation of the material occurs as the knife penetrates. Which category a specific size reduction task falls into depends essentially on the material behavior, the loading rate — i.e., in the case of the cutting mill, the rotor speed — and the geometry of the cutting edges. The influence of geometric conditions on the development of tensile stresses has already been described. High tensile stresses lead, in the presence of a crack, to a high crack propagation rate and thus to brittle fracture [Wol04, p. 11]. Brittle materials such as thermosets or, among thermoplastics, polystyrene or ABS tend to fracture in a brittle manner in the cutting mill; at high loading rates, brittle fracture behavior of plastics generally also occurs. Time-dependent relaxation processes cannot take place to a sufficient extent, so that the high stresses lead to the fracture limit of the material being exceeded.

Overview of Influencing Variables on the Size Reduction Process

One objective of size reduction is the production of the most homogeneous possible ground material with a narrow particle size distribution at high plant throughput and low energy input. The choice of the appropriate machine is made according to the material, the size of the feed material, and the particle size to be achieved. In general, the following broad trends regarding material-related, operational, and design-related influencing variables on the size reduction process can be identified.

Material-Related Influencing Variables

Ductile material tends to require more energy for size reduction than brittle material, because a significant proportion of the energy input must be expended on deformation. For brittle materials, on the other hand, less energy is required, because the crack front propagates suddenly without further energy input. When the appropriate size reduction machine is selected, narrower particle size distributions tend to be achieved with ductile materials, because the material is separated in a defined manner and does not break uncontrollably into fragments of varying sizes with a high fines content, as is the case with brittle materials [Höf86, p. 47 f.].

Operational Influencing Variables

As already described, high peripheral speeds lead to brittle material behavior. More energy tends to be required in this case, because the fracture stress of the material is higher. With impact or percussion stress, however, the energy required decreases, because the very high loading rates of brittle materials lead to fracture with unstable crack propagation. A rather broad particle size distribution results for the same reasons as with materials that are brittle from the outset. The higher speed leads to faster size reduction and thus to higher plant throughputs [Höf86, p. 47].

Design-Related Influencing Variables

In general, the required drive power increases with the number of rotors and the width of the rotors, but at the same time the throughput also increases. A higher number of rotor and stator blades leads to more frequent stressing of the material and thus to higher throughputs. Smaller wedge angles under shearing and cutting stress tend to result in lower energy consumption, but cause high wear on the cutting edges. Larger gap widths produce higher bending stresses. For ductile materials, this leads to greater plastic deformation and thus to higher energy requirements. In systems with a screen bottom, reducing the screen hole size means the material is stressed more frequently before it can leave the grinding chamber. This reduces throughput and increases the energy required [Bau01, p. 37; Höf86, p. 48].

Comminution of Fiber-Reinforced Thermoplastic Polymers

Fiber-reinforced plastics have been experiencing increasing demand for many years due to their good weight-specific mechanical properties. In particular, continuous fiber-reinforced thermoplastic semi-finished products, known as organo sheets, are increasingly being used in the automotive sector to save weight. Consequently, it can be assumed that in the future there will be an increasing amount of plastic waste containing glass or carbon fibers. Due to the original use of the continuous fibers as a fabric, comminuted waste of this kind can no longer be returned to this state. Therefore, these materials can only be processed into short fiber-reinforced recyclate in the sense of material recycling. It should be noted that the comminution of the fibers leads to a reduction in mechanical properties.

During the comminution process, particular attention must be paid to the generation of fiber dust. Fiber fragments that are produced represent a potential health risk because they may be inhalable and respirable. So-called WHO fibers, whose length is >5 mm, whose diameter is <3 mm and whose length-to-diameter ratio is >3:1, are considered particularly hazardous to health.

In the following, research results from Kunststofftechnik Paderborn at the University of Paderborn are presented, which were compiled and published by Tölle and Hopp et al. [THB+22; TH21; TH22a; TH22b; THH21; TMR+22]. Previous studies have dealt with the influence of the comminution process on the mechanical properties of the recyclate produced from the ground material, or with increasing throughput or minimizing energy consumption during the comminution process. However, investigations into the influencing factors on fiber dust generation have not previously been carried out in this form.

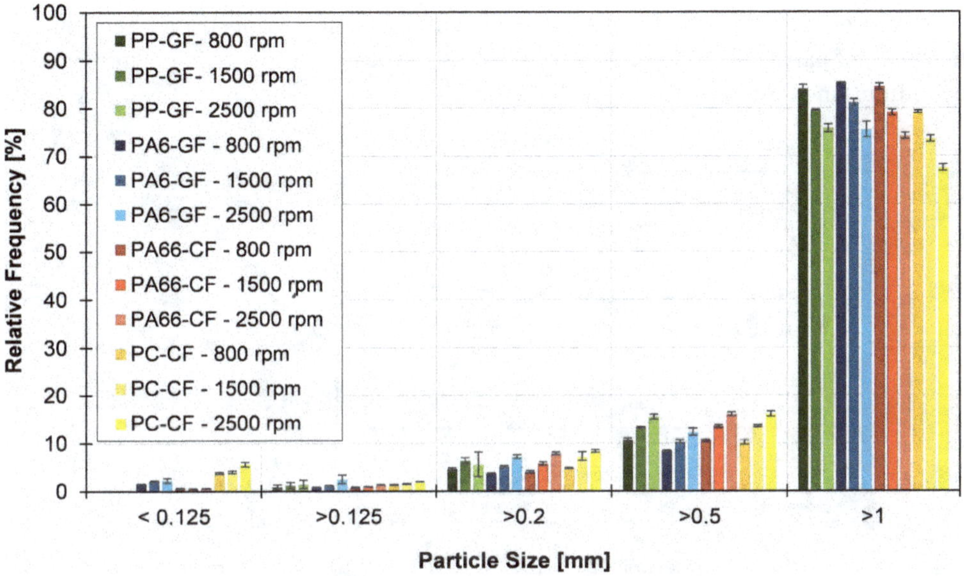

Figure 3.16 Particle size distribution of dust from comminuted, continuous fiber-reinforced thermoplastic semi-finished products by a cutting mill as a function of material and speed

Figure 3.16 shows the particle size distribution of fiber dust (particles <1 mm) from comminuted organo sheets with different material combinations as a function of the rotor speed of the cutting mill used (SM300 from Retsch GmbH). Glass fiber-reinforced organo sheets made of polypropylene (PP, Tepex® dynalite 104-RG600(x)/47%, 1 mm thickness) and polyamide 6 (PA6, Tepex® dynalite 102-RG600(x)/47%, 2 mm thickness) and carbon fiber-reinforced organo sheets made of polycarbonate (Tepex® dynalite 210fr-C221(x)/45%, 1 mm thickness) and polyamide 66 (Tepex® dynalite 201-C200[x]/50%, 7 mm thickness) from Bond Laminates GmbH were used. For the comminution process, the feed size is 30 x 30 mm, the screen mesh size is 10 mm and the throughput is 50 g/1x. The results show that an increase in speed produces a greater proportion of smaller particles for all materials. As previously stated, this is due to the time-dependent viscoelastic material behavior of the plastics, in which an increase in the rate of stress loading leads to more brittle fracture behavior. Furthermore, the organo sheets also show different comminution results among themselves, which particularly concerns the particle fraction <0.125 mm. The polycarbonate material with carbon fibers contains the most particles of this size, while the ground material of the polypropylene organo sheet shows no particles in this fraction. This is due to the more ductile material behavior of PP compared to PC. The PC fractures in a brittle manner, which leads to the formation of fragments of varying sizes and in particular to a larger fines content. For the polyamide materials with fiber reinforcement, brittle fracture behavior is also to be expected, which is also demonstrated here by the detection of particles <125 µm. The results thus support the influencing factors on the comminution result presented in this chapter.

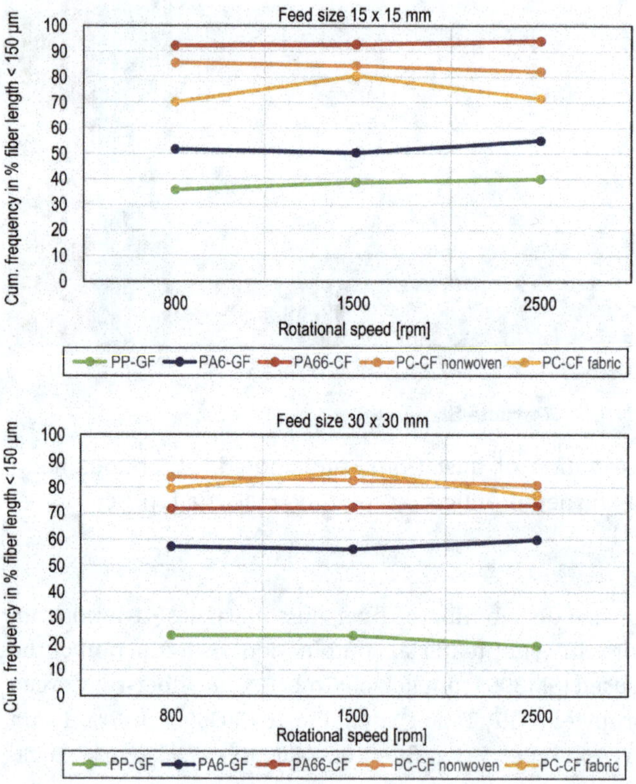

Figure 3.17 Fiber length distribution of the dust fraction <125 µm as a function of material, feed size and speed

By pyrolysis of the matrix material, the fiber length distribution for the respective smallest dust fraction is evaluated using a scanner and the evaluation software FiVer (SKZ – KFE gGmbH). The cumulative frequency of fiber lengths <150 µm for the organo sheet samples as a function of feed size and rotor speed is shown in Figure 3.17. First, it becomes clear that the speed has no discernible influence on the formation of fiber fragments in this fiber length fraction. Thus, although smaller particles are produced in the ground material at high speed, no further shortening of the fibers contained in the particles occurs within the size classes investigated. Based on the results for polypropylene and polyamide 66, it can be seen that reducing the feed size for these materials leads to a significant increase in short fibers. With the same mass-related feed quantity (50 g/1x), a smaller feed size results in a greater number of collisions between the fed parts and the wall of the grinding chamber, which can explain this effect.

Further investigations into the generation of airborne dusts were carried out using a single-shaft shredder of the type Weima Smart Cutter WSC 250–400 from Weima Maschinenbau GmbH. Airborne inhalable fiber dust (dust <100 µm) was separated onto a cellulose nitrate filter using a VC25 collector. With a feed size of 250 x 250 mm, organo sheets were comminuted at a throughput of 1 kg/100 s over a period of 120 min. The dust deposited on the filter was weighed and, for comparability, normalized to the feed quantity in kg and the volume of air filtered during that period in m^3.

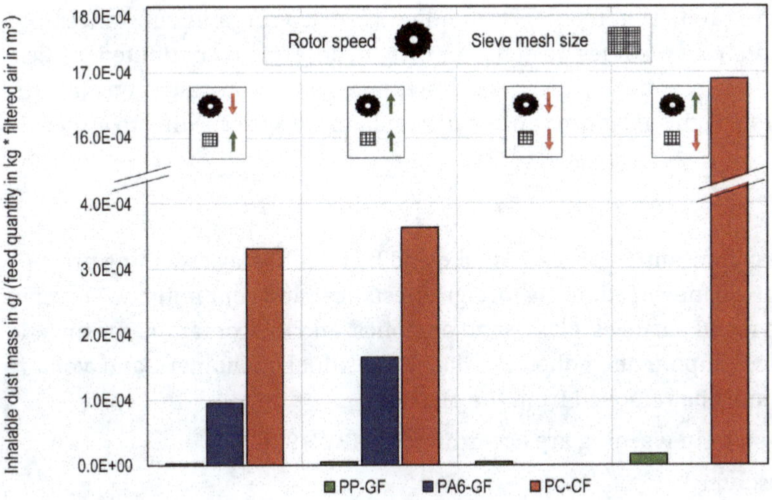

Figure 3.18 Normalized inhalable dust mass of airborne dust after comminution in a single-shaft shredder as a function of material and screen mesh size

Similar relationships can also be established for the airborne dusts as were shown in the sieve analysis. With higher speed, the dust concentration of airborne inhalable dusts also increases, which is shown for the PP, PA6 and PC organo sheets in Figure 3.18. Reducing the screen mesh size in the case of the brittle polycarbonate organo sheets with carbon fiber reinforcement leads to a 4.5-fold increase in the normalized E-dust mass. This can be explained by the higher number of cutting operations required before the material can leave the grinding chamber. This effect increases the fines content in the ground material and also the airborne dust quantity.

In collaboration with the Institute for Prevention and Occupational Medicine IPA of the German Social Accident Insurance, initial assessments of health hazards have been made on the basis of the comminution investigations and the collected fiber dusts. In the fiber dust of the carbon fiber-reinforced organo sheets, carbon fiber fragments are detectable that meet the WHO characteristics. Appropriate respiratory protection must be observed here. However, the tests carried out show only minor cytotoxic effects, which is why the health risk is not to be classified as higher than that of biopersistent, non-fibrous dusts [TMR+22].

3.2.2 Washing, Separation and Drying

Comminution produces a ground material bulk with a more or less narrow particle size distribution. A not insignificant fines content is often still present. The aim of the subsequent process steps is to free the ground material from foreign substances such as metal, stones and sand, from further contaminants such as organic substances and adhesive labels, and from the fines content. For this purpose, the comminuted particle bulk is washed and subsequently dried. Intermediate separation stages ensure further separation of foreign substances and in particular sorting according to a specific plastic fraction.

Washing

In most cases, the comminution process is followed by a multi-stage washing process. The plastic waste is cleaned of adhering foreign substances and contaminants. Foreign substances can be metal particles, glass, sand or stones, and the contaminants include adhering biogenic components, adhesives, labels, residual monomers and volatile components that must be removed from the plastic.

The essential sub-steps in washing are according to [MBW+92]:

- Soaking or dissolving
- Detachment of the contamination
- Separation of the contamination

The washing systems differ considerably from one another depending on the cleaning task. The required washing systems, often referred to as washing modules, are appropriately integrated into a recycling line and address the above-mentioned sub-steps either in a single washing module or in systems connected in series.

The following systems are commonly used today for cleaning plastic waste:

- Pre-washing unit with multiple separation chambers
- Wet cutting mill

- Friction washer
- Hot wash

Other designs are:

- Washing drum
- Stirred tank
- Screw conveyor

These systems are combined with intermediate separation stages, settling tanks, contaminant separators, and elaborate process water treatment. In the following, the four first-mentioned individual systems will be discussed in more detail.

In pre-wash units pre-shredded plastic waste is freed from contaminants such as metal, stone, glass, or sand in order to protect downstream machines from damage. In addition, soaking of other impurities and an initial washing of the waste take place. For this reason, this type of washing module is placed at the beginning of the cleaning process. Figure 3.19 shows schematically the design of such a pre-wash unit of the type manufactured by Herbold Meckesheim GmbH.

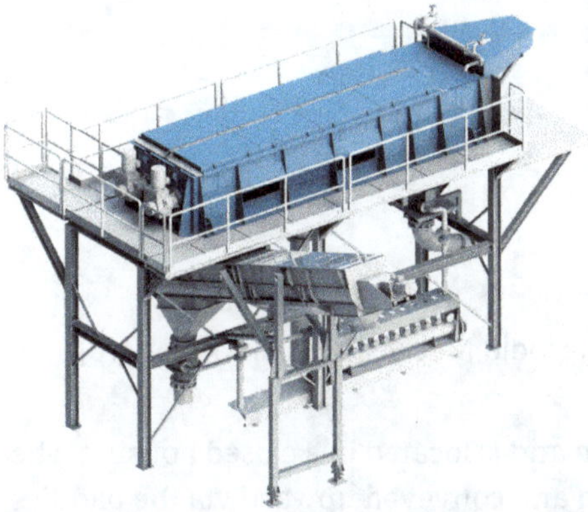

Figure 3.19 Schematic design of a pre-wash unit with separation chambers connected in series, slightly modified after [NN06]

Basically, the system consists of several separation chambers in which heavy contaminants can be separated. The material is conveyed into the separation chambers by means of conveying screws and at the end reaches the next washing unit via a discharge screw. In the first process stage, the feed material enters the feed chute via a conveyor belt or a conveying screw, which is arranged above the first pre-wash tank. Heavy foreign matter such as stones sinks to the bottom, while the comparatively light product material floats to the surface and is conveyed into the next separation chamber. To prevent product material contaminated with impurities from sinking, a counter-directed flow provides the necessary buoyancy. In the second stage, washing nozzles are arranged that loosen lightly adhering impurities such as sand, dirt, paper, or biogenic contaminants. Part of the contamination is flushed through the screen bottom arranged below the feed material. The other part is conveyed together with the product material into the next separation chamber. There, the foreign matter sinks to the bottom, while the product material is conveyed via the discharge screw into the next system module. The sink material is discharged from the system via double air-locks or screw systems [NN08].

The wet granulator provides fine size reduction of film waste into film flakes or of pre-shredded PET bottles into PET flakes after pre-shredding, for example in a single- or multi-shaft shredder. During the size reduction process, which in principle functions like a dry-operated granulator, large quantities of water are added to the cutting chamber. The stresses prevailing in the granulator, in combination with the water, cause intensive cleaning of the surfaces. In addition, the plastic material is treated gently and the heat generated is dissipated. Typical wet granulators operate at rotor speeds between 450 and 600 rpm and throughput capacities of up to 3000 kg/h [NN09-ol].

The friction washer is used for cleaning and dewatering of plastic waste. It is sometimes used as a downstream unit of a wet granulator.

Figure 3.20 Friction washer FA-60–300 from Herbold Meckesheim GmbH [NN10]

An inclined paddle rotor is enclosed by a screen and is located in a closed housing. The feed material with water is fed in at the bottom and conveyed upward via the paddles mounted at an angle on the shaft (see Figure 3.20). In the process, cleaning of the plastic regrind takes place through the shear input of the paddles, while the water with the foreign matter escapes through the screen via the centrifugal forces generated by the rotation. The dewatering of the friction washer is also used to exchange the process water and to transfer the material to a further cleaning or separation stage [NN11].

To remove strongly adhering residues, glued-on labels, organic contaminants (e.g., cooking oil), or odorous substances, hot washing is frequently employed. Depending on the requirements, different washing media are used with the addition of surfactants. Since the operating costs of hot washing systems are comparatively high due to the elevated temperature of the washing medium and the use of caustic solutions, they are used for the production of higher-quality recyclates, e.g., for the food sector. Thus, they are used in recycling lines for PET bottle recycling as a preliminary stage for even more intensive cleaning steps (e.g., the URRC process). In the meantime, the systems have also been established for further material streams, such as polyolefin fractions, which are contaminated by greasy impurities such as body lotion or sunscreen [NN12-ol; NN13-ol].

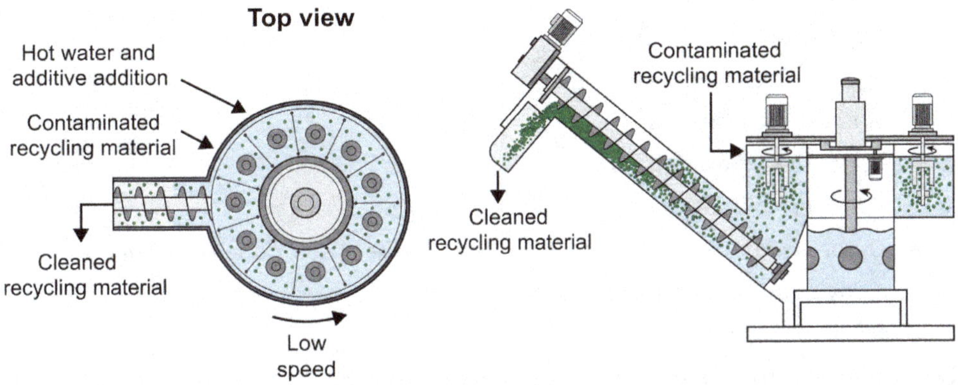

Figure 3.21 Schematic design of a hot washing system, slightly modified after [NN07-ol]

As can be seen from Figure 3.21, such a system consists of a washing tank that is rotatably mounted and rotates slowly. Around the axis of the tank, individually driven, fast-rotating rotor washers are arranged, which ensure the movement of the regrind to be washed in the washing medium and the detachment of heavy soiling. A discharge screw conveys the product material out of the hot washing unit. The recycling equipment manufacturer Lindner Washtech has developed a three-stage hot washing process which, in addition to the hot washing tank equipped with mixing arms and a friction washer, incorporates a washing unit in between that is designed similarly to the pre-wash unit. Here, the material is subjected to constant friction and adhesions are gently removed [NN14-ol].

Separation

After the shredded plastic waste has been cleaned, separation takes place according to plastic type or color, whereby additional separation of foreign matter that may still be present in the material bulk can occur. For this purpose, various physical or optical methods are used, which are also already applied in the sorting of waste in a light packaging sorting plant, such as NIR sorting. In contrast to the sorting of non-shredded waste pieces, the sorting of the feed material after shredding and after the washing process can be carried out considerably more accurately. Overlapping waste pieces, adhering impurities, or composites that have not yet been broken up no longer play a role. Each individual particle can, depending on the degree of size reduction and the possibility of singulating the particles, be detected and separated according to the respective plastic type.

In principle, the following separation techniques are suitable for separating the regrind:

- Density separation
  - Float-sink separation
  - Jigging process, tabling
  - Hydrocurrent separator
  - Air current separator
- Sensor-based separation
  - NIR spectroscopy
  - Optoelectronic sensors, for example by means of color line-scan camera
  - X-ray fluorescence sensors
- Electrostatic separation

This list makes no claim to completeness, but is intended only to provide an overview of methods frequently applied to plastic waste. Further methods are described, for example, by Martens et al. [MG16, p. 286 ff.].

Density Separation

The separation or sorting of regrind using density differences is a method that works very well for the separation of heavy foreign matter, but also plays a major role in separation by plastic type due to density differences. Separation is achieved by the different behavior of the particles in the regrind in interaction with the surrounding fluid (liquid or air). If the fluid is not at rest but flows within the apparatus, then the size and shape of the particles also play a role. Therefore, the production of a regrind with a narrow particle size distribution is advantageous for such apparatus [MG16, p. 45].

As can be seen from Figure 3.22, for certain mixed fractions, such as the PET bottle fraction with HDPE or PP screw caps, there are clear differences in density. The separation of PA from the polyolefins PE or PP is also possible using water as the separation liquid. For other fractions, such as polyamide and PET, density separation can be carried out with separation liquids of higher density; for PE and PP, however, separation from each other by density is not possible.

In practice, challenges arise from the use of additives in plastics. Density-modifying fillers can cause the density to increase, making separation impossible. On the other hand, the density decreases significantly through the foaming of plastics, so that, for example, polystyrene with a higher compact density than water floats as a foamed variant.

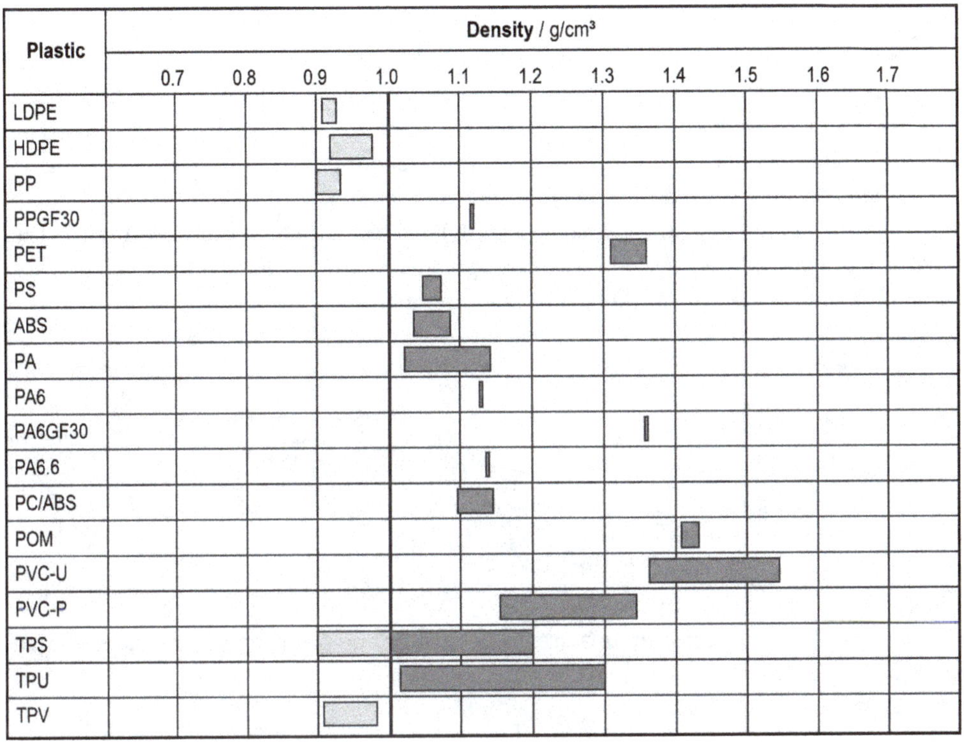

Figure 3.22 Density ranges for various plastics [BBO+13; Eng15; NN15-ol]

The principle of density separation in a gravity separator is based on a simple physical principle. The particles are placed in a liquid of a specific density; in the case of water, this is approximately 0.998 g/cm^{3}at a temperature of 20 °C. The particles in the water with a lower density remain at the water surface (float material), because the buoyancy force, which corresponds to the weight force of the displaced water volume, is greater than the opposing weight force of the particles. The particles with a higher density sink downward in the water accordingly (sink material). In addition to water, a salt solution or a water-alcohol mixture can be used to adjust the density. The particles to be separated should be between 5 and 10 mm in size and wettable (no air bubble adhesion). In Figure 3.23, left side, a gravity separator is shown schematically. Paddle movements create a gentle, superficial flow in order to continuously skim off the float material. The sink material is discharged at the bottom of the tank via airlocks or dewatering screws. Further designs are described in [Sch96, pp. 18–26]. To prevent the sink material from being swept up, turbulence must be avoided at all costs. However, this risk considerably limits the throughput of these systems. Gravity separators operate only with gravitational force, i.e., with a mean acceleration of 9.81 m/s^2, and the influence of grain size and shape cannot be neglected [MG16, p. 47; BH13, p. 518; BLS+17, p. 450].

For higher throughputs and better separation sharpness, sorting centrifuges can be used (see Figure 3.23, right side). These consist of a drum that tapers conically at both ends. Inside the drum, a screw following the contour is arranged, whose flights are divided into two parts and allow a conveying effect toward the respective ends. The previously homogeneously mixed suspension of separating liquid and plastic particles enters the central area of the centrifuge via a hollow shaft. The centrifugal force presses the suspension against the drum wall. Particles with a higher density than the separating medium are pressed more strongly against the wall than lighter particles, which float on the surface of the liquid from the inside.

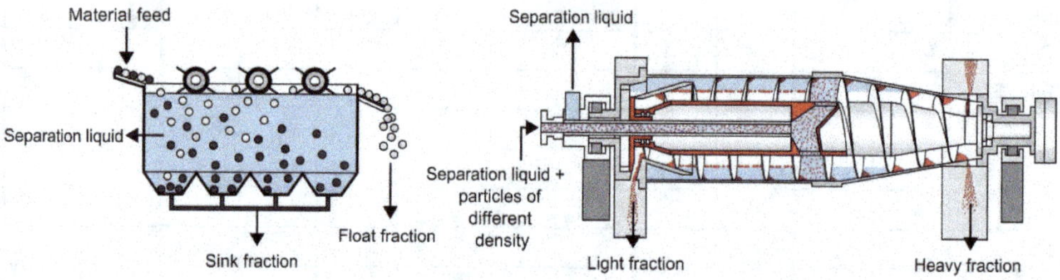

Figure 3.23 Schematic design of a gravity separator and a sorting centrifuge, adapted from [MG16, p. 48]

The screws each provide a conveying effect for the heavy and light fractions out of the centrifuge. The conical outlet of the drum at the ends enables dewatering of the suspension. The separating liquid exits the centrifuge at the end where the light fraction is discharged and is returned to the circuit. With such a system, plastics that are close together in density can be separated from one another. The high peripheral speeds of up to 4000 rpm generate high acceleration values of up to 1100 g acting on the particles. In addition, high shear forces exist between the particles and the liquid, which produce a good cleaning effect. The disadvantages of this system lie in the complexity of the process engineering and the high operating and investment costs [KC10, p. 173, BLS+17].

A more cost-effective alternative without major maintenance requirements is the hydrocyclone. Here there are no moving plant components that lead to wear during operation and consequently to tight maintenance and servicing intervals. As can be seen from Figure 3.24, the feed material, consisting of liquid and the particles to be separated, is introduced tangentially at high velocity into the cyclone. Due to gravity and the shape of the cyclone, a downward-directed so-called primary vortex forms. In the centrifugal field, the heavy particles are pressed against the cyclone wall, while the light particles are carried toward the center of the cyclone. Due to the conical inlet of the cyclone, the lower opening acts as a throttle and only allows the heavy particles near the wall to pass through.

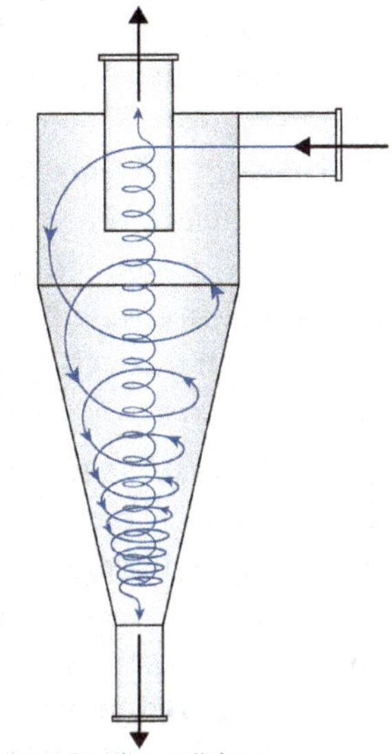

Figure 3.24 Schematic design of a hydrocyclone (left) and a cylindrical centrifugal force separator (right), adapted from [MG16, p. 48; SBL21, p. 8]

The particles and the liquid in the center of the cyclone are forced to reverse direction by the flow conditions, whereby an upward-directed so-called secondary vortex forms, which leads into the upper outlet of the cyclone and results in the discharge of the light fraction. The particles are subjected to an acceleration of up to 100 g. Using a hydrocyclone, the polyolefin fraction (PP/PE) can be recovered with a separation efficiency of approx. 95–98 %.

To improve the separation result, hydrocyclones can be connected in series, whereby downstream separation stages can be operated with a salt solution in order to discharge, for example, polystyrene with a density of 1.05 g/cm^3 as the light fraction [KC10, p. 171, BH13, p. 519, BLS+17, p. 450].

More recent developments aim at the use of cylindrical centrifugal force separators, whose axis is arranged at an angle in space (see Figure 3.24 right). These devices, previously known only from coal or mineral raw material processing, have the advantage that the feed material does not have to be introduced into the separator together with the separating medium. This allows blockages and wear on the pumps to be avoided. The separating medium is introduced tangentially at the lower end of the separator and discharged tangentially again at the upper end. This creates a centrifugal field with an air column in the interior. The feed material enters the center of the vortex at the top and is captured by it. Particles with a higher density than the separating medium are pushed outward and discharged from the separator via the upper outlet together with the separating medium. The light particles remain inside at the interface with the air column and exit the separator through the lower end due to the inclination of the separator.

The advantage of the cylindrical design results from the better realization of a longer residence time in the centrifugal field, which in combination with the settling time of the feed material in the gravitational field is advantageous for the separation of density classes. In the case of conically tapering sorting cyclones, a high inlet velocity is necessary to maintain the centrifugal force field. However, this is at the expense of residence time. With cylindrical centrifugal force separators, the residence time can be easily increased by extending the length of the cylinder. A further advantage results from a lower pressure drop with this design. In the conical variant, a strong flow reversal occurs in the cyclone, which leads to high pressure drops when throughputs are increased [BLS+17, p. 451].

A dry variant of density separation is represented by air classifier sorters, one representative of which, the so-called zigzag classifier, is shown in Figure 3.25. The classifier operates on the counterflow principle, in which air flows from bottom to top and the feed material consisting of particles of different densities is fed into the classifier from above. The heavy fraction is separated downward, while the light fraction can be discharged upward by the air flow. Due to the design of the classifier, the feed material comes into contact with the air stream multiple times, whereby cross-flow classification can take place. Each section thus represents a separation stage in which an air vortex is created by the air flow in interaction with the limiting classifier geometry.

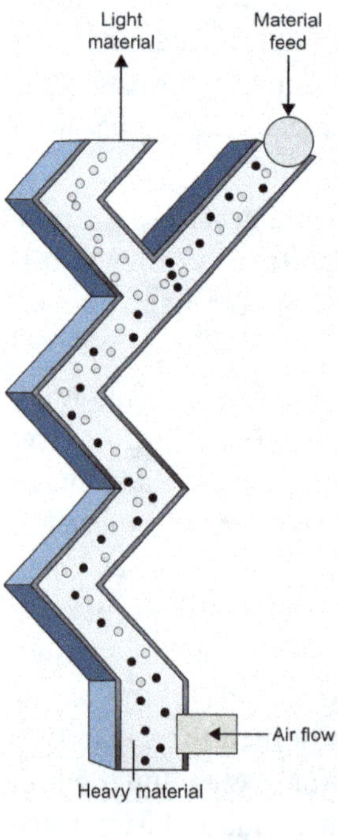

Figure 3.25 Schematic design of a zigzag air classifier, after [MG16, p. 49]

The coarse material deposits at the inner edge on the horizontal flank in each case. The light fraction is carried upward from stage to stage and separated via a cyclone and a rotary valve. To increase throughput, multi-channel classifiers can be used, which represent a simple parallel arrangement of several classifier channels side by side [MG16, p. 49; Sti09, p. 314 f.; NN16-ol; NN17-ol].

Sensor-Based Separation

If plastics cannot be separated by density, or only with great effort, sensor-based systems can be used for separation by plastic type or also by color. For this purpose, the NIR spectroscopy already introduced in section 3.1.7 is suitable, which can achieve a significantly finer separation of the feed material after the comminution and washing process than is possible in a light packaging sorting plant. Furthermore, optoelectronic sensors, such as color line-scan cameras, are used for color detection, or X-ray fluorescence sensors for the detection of chloride-containing plastics such as PVC [MG16, p. 59].

Electrostatic Separation

The sorting of particles in an electric field aims at a differing deflection force on the particles in the electric field. This is essentially dependent on the Coulomb force and gravity. To achieve a Coulomb force in the electric field, the particles must carry charges of a certain magnitude and polarity. In the case of plastics, this is achieved by triboelectric charging in mixing drums or in fluidized beds (see Figure 3.26). Through friction, a charge exchange takes place, whereby one of the plastic types to be separated is charged positively and the other negatively. If the particles then fall in a free-fall separator between two plate electrodes, the Coulomb force acts on the particles and deflection of the particles toward one of the electrodes occurs. In this way, the separation of the two plastic types is achieved [Sch96, p. 222; MG16, p. 54].

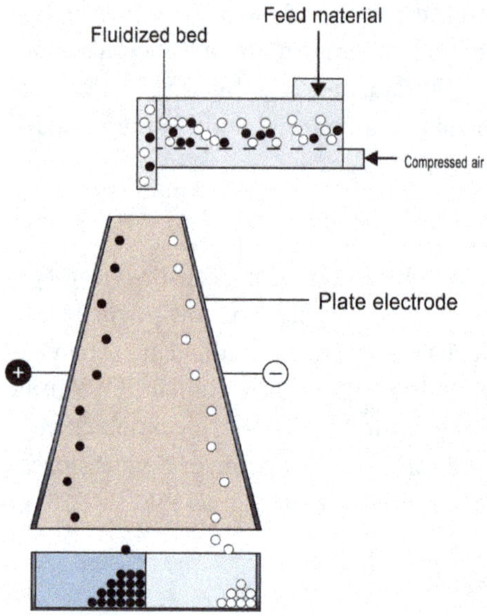

Figure 3.26 Schematic design of an electrostatic free-fall separator with triboelectric charging, after [MG16, p. 53]

According to their molecular structure, plastics charge up differently. Functional groups serve as electron pair donors or acceptors. So-called charge series provide information on whether different polarities can be achieved during triboelectric charging or not. An excerpt from such a charge series is shown in Figure 3.27. A plastic located further down in the table would charge negatively through friction with a plastic located further up.

Charge tendency	Plastic	Functional Group
+	PUR	-N-CO_2
↑	PC	-O-CO_2
	PA	-N-CO
	PS	
	PE	
	PP	
↓	PET	
-	PVC	-Cl

Figure 3.27 Triboelectric series, according to [MG16, p. 54]

The separation of PVC from polyolefin fractions or PET is one example of the application of this technology. The separation of PP and PE can be achieved by adding surface-active substances such as nigrosine [Sch00, SNS+04], by electron irradiation [HPS+09] or by plasma pretreatment [HPS+09, SNS+04]. Good separation results can also be achieved for the separation of ABS and PS [PS13].

With plastic waste, it must be taken into account that it is not solely the respective polymer that possesses specific triboelectric properties due to its molecular structure.

Depending on the previous application, the plastics contain a wide variety of fillers. In this context, dyes, carbon blacks and chalk have little influence on the electrostatic charging of the plastics, while talc always leads to negative charging, as results from corresponding investigations within the framework of a research project at TU Freiberg in cooperation with the Leibniz Institute of Polymer Research Dresden e. V. show [PS13].

Drying

Depending on the material stream to be processed, the shredded plastic particles are either dried directly after washing and further processed, or separation by plastic type (density sorting using water as the separation medium) takes place in the wet state. This is followed by the drying process, which is urgently required for subsequent storage and, in particular, for further processing in screw machines (extruders, plasticating units). If moisture is present in the melt during processing, undesirable hydrolytic material degradation occurs, for example with PET or PA, and, if the moisture cannot escape via degassing, voids form in the component.

With non-hygroscopic plastics such as PE or PP, the liquid adheres only to the surface, whereas hygroscopic plastics such as PET or PA can absorb water into their molecular structure. With regard to surface adhesion, the geometry, i.e. the shape and size of the particles to be dried, plays a significant role in the total amount of adhering liquid. Very thin packaging films, for example, have a very high surface-to-volume ratio as shredded film flakes, whereas this is lower for shredded shampoo or beverage bottles. This must be taken into account when designing the drying line.

Available options are mechanical drying, in which separation from the liquid phase is achieved by external force in combination with retaining screens, and thermal drying, in which a hot air stream provides the necessary dehumidification.

Purely mechanical drying is carried out via inertial forces or induced field forces in combination with screens. Dewatering screws are used for coarse separation, conveying the plastic particles out of the washing processes, for example. An inclined unit, mostly open at the top, contains a conveying screw on the inside. The conveying trough consists of a screen base, allowing the material being dewatered to drip off. Friction washers have a similar design, but must primarily fulfil a washing function. For this purpose, they operate at higher speeds and dewatering takes place via centrifugal forces over the entire lateral surface [BBM+95, p. 226].

After coarse separation, centrifuges of various designs can be used as a second drying stage. Due to the residual water in the feed material and the high peripheral speed, these dryers exhibit a certain cleaning effect in addition to dewatering. Dust and other foreign matter still adhering to the plastic particles can be removed along with the water. Horizontal centrifugal dryers contain a screen basket surrounding the rotor on the inside, through which dewatering is carried out. Depending on the type, so-called fan wheels are installed on the rotor at the material inlet and outlet, which are intended to create suction at the inlet and ejection at the outlet and to prevent blockages.

Another design is represented by stage dryers, which are built vertically and have several drying stages. The feed material enters the dryer at the bottom and is transported upward in the vertical direction via transport paddles. The rotor diameter increases from bottom to top, so that the peripheral speeds also increase. In the uppermost stage, the maximum peripheral speed ensures residual moisture levels in the material of 0.4 to 0.06 wt.-% [BBM+95, p. 227; NN19, p. 1].

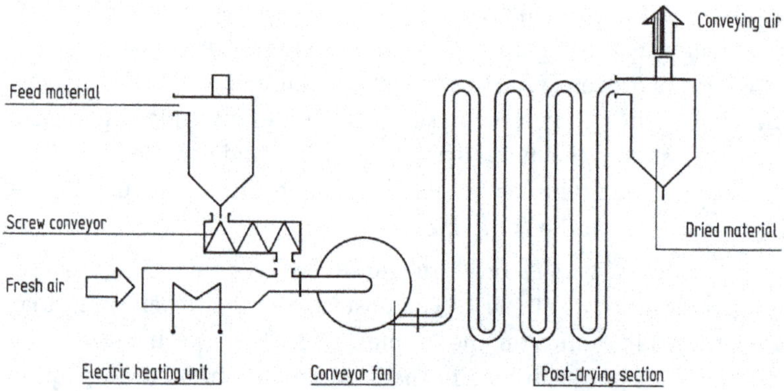

Figure 3.28 Schematic representation of thermal drying [BBM+95, p. 228]

In thermal dryers, a hot air stream is used to remove the moisture contained in the feed material. The feed material is fed, for example via a screw conveyor, into the heated air stream and guided through a drying section in a pipe coil, in which the warm air removes the moisture adhering to the feed material (see Figure 3.28). A cyclone at the end of this drying section separates the feed material. The pipe coil can have a length of approx. 30 m in order to ensure the required residence time. In this way, residual moisture levels of less than 1 % are achieved, which is not sufficient for further processing in an extruder or injection molding machine. Further drying can then be carried out using dry air dryers commonly used in plastics processing.

3.2.3 Regranulation and Agglomeration

At the end of the processing stage, a free-flowing recyclate bulk material is produced, either in granulate form or as a so-called agglomerate. For granulation or, in the case of recyclate production, regranulation, the shredded regrind or film flakes are first plasticated in a screw machine. This is frequently done using a co-rotating twin-screw extruder, in which the material can simultaneously be compounded by adding additives. Granulate is then produced by the following two process variants:

- Hot-die-face cutting
- Cold granulation

In the hot-die-face cutting process, the plastic melt is forced through a die plate, which has a varying number of holes depending on the throughput and extruder size. A rotating knife is arranged at the exit of the holes, which scrapes the material off the die plate (see Figure 3.29, left side). The cooling of the cut, still molten plastic particles is carried out either by air or by a water stream that flows around the rotating knife and, in addition to cooling, has the task of transporting the granulate away from the knife. Due to the stored heat and the associated possibility of further deformation, the plastic particles tend toward a spherical shape upon contact with the water. The spherical or lenticular shape of the granulates produced in this way (see Figure 3.29 right) has advantages in terms of flowability compared to the cylindrical granulate shape produced during cold granulation.

The second process variant, cold granulation or strand granulation, takes place in two steps. Here too, the melt first passes through a die plate, but the strands are then guided through a water bath in which they cool down. The strands are subsequently cut into the typical cylindrical granulate shape in a granulator (see Figure 3.30). The cutting process produces burrs at the cut edges, which can cause the granulate particles to interlock during flow processes.

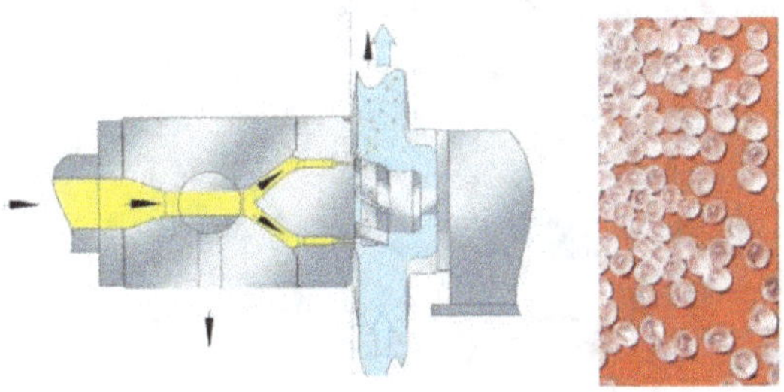

Figure 3.29 Principle of underwater granulation (left) and lenticular granulate (right) [Sch11, p. 95]

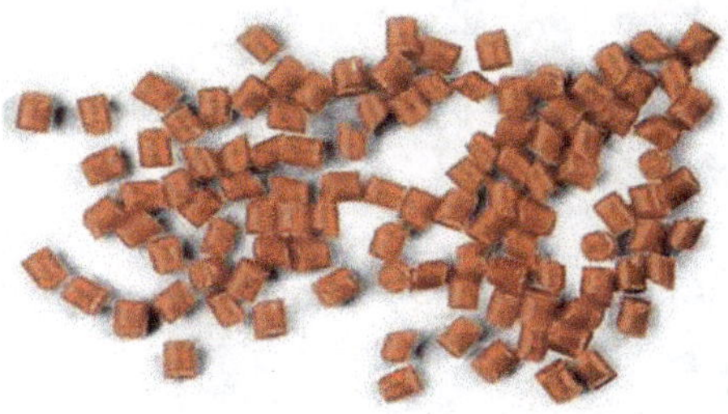

Figure 3.30 Granulate in cylindrical form [Sch11, p. 96]

In addition to granulate production, agglomeration is also used to produce the flowability of the regrind, particularly for feed materials with a low bulk density such as films, foams or fibers. There are various designs of agglomerators for plastic recyclates, of which the system from Wipa Werkzeuge & Maschinenbau GmbH is presented below as a representative example (see Figure 3.31). The system shown consists of the feed screw on the left, fed via a hopper, the pair of discs housed in a casing (visible in red in the center) and the blower with the separation cyclone visible in the background. For trouble-free continuous operation, a buffer container equipped with an agitator to prevent bridging is used instead of the hopper.

A feed screw compacts and conveys the feed material between the pair of discs. This consists of a stationary stator disc, through the center of which the material is conveyed, and a rotor disc, which features radially arranged grooves at an angle and bolted-on kneading bars. In the disc gap, kneading and rolling stresses act on the material, leading to heating by friction.

The plastic is partially melted and finger-sized agglomerates are formed (see Figure 3.32). These are forced outward from the disc pair by centrifugal forces and by the advancing material flow, and are separated via the blower and cyclone. A downstream cutting mill reduces the agglomerate to the desired particle size.

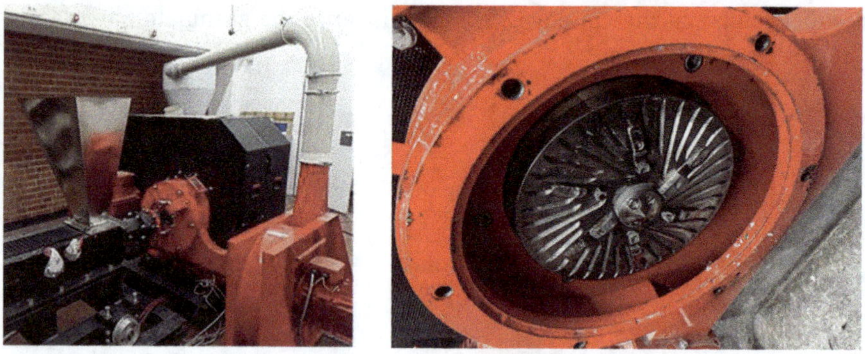

Figure 3.31 Agglomerator (left) and view of the rotor disc (right) of an agglomerator from Wipa Werkzeuge & Maschinenbau GmbH

Figure 3.32 Production of agglomerate from shredded LDPE film material

References

[Bau01] Bauer, W.: „Untersuchung des Einzelzerkleinerungsvorganges in Schneidmühlen am Beispiel von Polypropylen". Fortschrittsreihe VDI-Reihe 3, Nr. 694, VDI-Verlag, Düsseldorf 2001

[BBM+95] Brandrup, J.; Bittner, M.; Michaeli, W.; Menges, G.: „Die Wiederverwertung von Kunststoffen". Carl Hanser Verlag, München, Berlin, 1995

[BBO+13] Baur, E.; Brinkmann, S.; Osswald, T.; Rudolph, N.; Schmachtenberg, E.: „Saechtling Kunststoff Taschenbuch". Carl Hanser Verlag, 31. Ausgabe, München 2013

[Bee17] Beel, H.: „Sortierung von schwarzen Kunststoffen nach ihrer Polymerklasse mit Hyperspectral-Imaging-Technologie". In: Recycling und Rohstoffe, Hrsg. Thomé-Kozmiensky, K. J.; Goldmann, D., Band 10, TK-Verlag, Neuruppin, 2017

[BH13] Bilitewski, B.; Härdtle, G.: „Abfallwirtschaft". Springer-Verlag, Berlin, Heidelberg, 2013

[BLS+17] Bauer, M.; Lehner, M.; Schwabl, D.; Flachberger, H.; Kranzinger, L.; Pomberger, R.; Hofer, W.: „Bestandsaufnahme und mögliche Perspektiven der nass-mechanischen Aufbereitung von Altkunststoffen für das rohstoffliche Recycling". In: Österreichische Wasser- und Abfallwirtschaft, Vol. 69, pp. 446–459, 2017

[BMW97] Bittner, M.; Marwick, J. von; Wolters, L.: „Aufbereitung von Kunststoffresten". In: Wolters, L.; Marvick, J. von; Regel, K.; Lackner, V.; Schäfer, B. (Hg.): „Kunststoff-Recycling. Grundlagen Verfahren –Praxisbeispiele". Hanser-Verlag, München, 1997

[CKF+20] Chen, K.; Kroell, N.; Feil, A.; Pretz, T.: „Determination of the composition of multilayer plastic packaging with NIR spectroscopy". Multidisciplinary Journal for Waste Resources & Residues, Vol. 13, pp. 62–66, 2020

[CKW+21] Chen, K.; Kroell, N.; Wickel, J.; Feil, A.: „Determining the composition of post-consumer flexible multilayer plastic packaging with near-infrared spectroscopy". Waste Management, Vol. 123, pp. 33–41, 2021

[DHM22] Dahlmann, R.; Haberstroh, E.; Menges, G.: „Menges Werkstoffkunde Kunststoffe". 7. vollständig neu bearbeitete Auflage, Carl Hanser Verlag, München, 2022

[Eng15] Engelke, N.: „Kunststoffrecycling: Effizient getrennt". In: Umweltmagazin, April-Mai, Springer VDI-Verlag, Düsseldorf, 2015

[GV18] Günther, S.; Vogt, M.: „Europas modernste LVP-Sortieranlage". Umweltmagazin, Juli-August, Springer-VDI-Verlag, Düsseldorf, 2018

[Hin71] Hinzpeter, A.: „Physik als Hilfswissenschaft – Teil 4: Elektrik". Vandenhoeck & Ruprecht, Göttingen, 1971

[Höf86] Höffl, Karl: „Zerkleinerungs- und Klassiermaschinen". Springer-Verlag, 1. Auflage, Berlin/Heidelberg/New York/Tokio, 1986

[HPS+09] Husemann, K.; Peuker, A.; Simon, F.; Reinsch, E.; Albrecht, V.: „Entwicklung und Optimierung von Verfahren zur Elektrosortierung praxisrelevanter Kunststoffabfallgemische". Schlussbericht zum AiF-Projekt, Vorhabennummer 14980 BR, 2009

[Jun04] Jungbauer, A.: „Recycling von Kunststoffen". Vogel Verlag und Druck KG, Würzburg, 1994

[KC10] Kranert, M.; Cord-Landwehr, K.: „Einführung in die Abfallwirtschaft". Wiesbaden, Vieweg+Teubner-Verlag, 2010

[KC10a] Kranert, M.; Cord-Landwehr, K.: „Einführung in die Abfallwirtschaft". Wiesbaden, Vieweg+ Teubner-Verlag, 2010, dort verwiesen auf: Anonym: Firmenschrift „Abfall-Trennungsanlagen". Nihot Luchttechnieken, Amsterdam

[KPB+17] Kreibe, S.; Pitschke, T.; Berkmüller, R.; Bokelmann, M.; Förster, A.; Stramm, C.; Pant, A.: „Umweltbezogene Bilanzierung von „intelligenten" und „aktiven" Verpackungen hinsichtlich der Recyclingfähigkeit und Durchführung eines Dialogs mit Akteuren der Entsorgungs- und Herstellungsbranchen". Umweltbundesamt, 2017

[LMK95] Löhr, K.; Melchiorre, M.; Kettemann, B.: „Aufbereitungstechnik. Recycling von Produktionsabfällen und Altprodukten". Hanser-Verlag, München, Wien, 1995

[MBW93] Michaeli, W.; Bittner, M.; Wolters, L.: „Stoffliches Kunststoff Recycling – Anlagen, Komponenten, Hersteller". Carl Hanser Verlag, München, 1993

[MG16] Martens, H.; Goldmann, G.: „Recyclingtechnik". Springer Fachmedien, Wiesbaden, 2016

[MBW+92] Michaeli, W.; Bittner, M.; Wolters, L.; Krebs, S.: „Materielles Recycling von Kunststoffen". Carl Hanser Verlag, München, Wien, 1992

[NGB+16] Nübler, D.; Gruna, R.; Brandt, C.; Küter, A.; Längle, T.; Kieninger, M.; Pohl, N.: „Innovative Technologies as Enabler for Sorting of Black Plastics". 19th World Conference on Non-Destructive Testing, 2016

[Nic96] Nickel, Werner (Hrsg.): „Recycling-Handbuch – Strategien – Technologien – Produkte". VDI Verlag, 1. Auflage, Düsseldorf, 1996

[NN01-ol] Thyssenkrupp Rassenstein GmbH: „Weißblechverpackungen mit 90,4 % Recyclingrate". Pressemitteilung vom 13.01.2020, https://www.thyssenkrupp-steel.com/media/content_1/presse/dokumente/2020_1/january_1/20200113pm_weissblechverpackungen_mit.pdf, abgerufen am 09.08.2022

[NN02-ol] Tomra Systems GmbH: „Mehr Intelligenz mit GAIN – Deep Learning basierende Add-On Lösung". https://www.tomra.com/en/sorting/recycling/tomra-technology/deep-learning, abgerufen am 11.03.2022

[NN03-ol] Steinert GmbH: „UniSort Black". Broschüre, https://steinertglobal.com/fileadmin/user_upload/_steinert/_downloads/_magnete-sensorsortierer/_sensorsortierung/UniSort_Black/UNI_black_DE_web.pdf, abgerufen am 11.03.2022

[NN04-ol] Inno-spec GmbH: „Hyperspectral Imaging in der Recyclingindustrie". https://inno-spec.de/recycling-industry/ abgerufen am 11.03.2022

[NN05-ol] UNTHA shredding technology: „RS Klasse – 4-Wellen-Zerkleinerer für industrielle Recyclinganwendungen". Produktbroschüre, 11/20, https://assets.ctfassets.net/ke8zgcubcfkc/6nEG3RFEd9mfDITysQIUqp/a3c20556ba17a60d6d3a49a0b334d9c4/RS50-150_12s_de_Web.pdf, abgerufen am 22.04.2022

[NN06] Herbold Mechesheim GmbH: „Vorwascheinheit VWE-Herbold-Meckesheim". Mit freundlicher Genehmigung zur Nutzung, 12.2025

[NN07-ol] B+B Anlagenbau GmbH: https://www.bub-anlagenbau.de/wp-content/uploads/2017/06/pet-hot-washer1.jpg und https://www.bub-anlagenbau.de/wp-content/uploads/2017/06/pet-hot-washer2.jpg, abgerufen am 25.04.2022

[NN08] Herbold Meckesheim GmbH: „Herbold Vorwascheinheit VWE". Infoblatt 58, Oktober 2019

[NN09-ol] Lindner Washtech GmbH: https://www.lindner-washtech.com/de/systemloesungen. abgerufen am 04.05.2022

[NN10] Herbold Meckesheim GmbH: „Friktionswäscher FA-60–300-Herbold-Meckesheim“. Mit freundlicher Genehmigung zur Nutzung, 12.2025

[NN11] Herbold Meckesheim GmbH: „Friktionswäscher Baureihe FA“. Infoblatt 46, Oktober 2019

[NN12-ol] „Massenanwendungen von Recyclaten ermöglichen“. Kunststoff Magazin, Pressemitteilung vom 24.05.2019, https://www.kunststoff-magazin.de/automatisierung/kunststoff recycling—massenanwendungen-von-recyclaten-ermoeglichen.htm, abgerufen am 04.05.2022

[NN13-ol] „Heißwaschanlagen“. Kuhn Fachverlag, Pressemitteilung vom 03. November 2019, https://kunststoff.kuhn-fachmedien.de/wp-content/themes/kuhn-magazin/pdf/mpdf/kuhn/print-post.php?postid=1889, abgerufen am 04.05.2022

[NN14-ol] Lindner Washtech GmbH, Pressemitteilung: „Lindner Washtech bringt neues Heißwaschsystem zur effektiven Kunststoffaufbereitung auf den Markt“.https://www.lindner.com/de/hotwashsystem, abgerufen am 04.05.2022

[NN15-ol] Kern GmbH Kunststoffwerke, Technische Datenblätter, https://www.kern.de/de/tech nische-datenblaetter-kunststoffe, abgerufen am 06.05.2022

[NN16-ol] JÖST GmbH + Co. KG: Produktbroschüre Zick-Zack-Windsichter, https://www.joest.com /wp-content/uploads/2022/05/2022_zick-zack-wind-sichter-de.pdf, abgerufen am 20.05.2022

[NN17-ol] Neue Herbold Maschinen- und Anlagenbau GmbH: Produktbroschüre Zick-Zack-Windsichter, https://neue-herbold.com/wp-content/uploads/2014/12/Zick-Zack-Sichter_ZS.pdf, abgerufen am 20.05.2022

[NN19] Herbold Meckesheim GmbH: „Mechanischer Trockner/Zentrifugaltrockner“, Infoblatt 49, Oktober 2019

[PS13] Peuker, A.; Simon, F.: „Elektrosortierung von Kunststoffen als kontinuierlich arbeitender Prozess am Beispiel der Trennung von PS/ABS“. Schlussbericht zum AiF-Projekt, Vorhabennummer 16531 BR, 2013

[SBL21] Schwabl, D.; Bauer, M.; Lehner, M.: „Advancing plastic recycling by wet-mechanical processing of mixed waste fractions“. In: Processes, Vol. 9, p. 493, 2021

[Sch96] Schubert, H.: „Aufbereitung fester Stoffe, Band II: Sortierprozesse“, Deutscher Verlag für Grundstoffindustrie, Stuttgart, 1996

[Sch00] Schultz, T.: „Untersuchungen zur elektrostatischen Sortierung der Kunststoffe Polypropylen und Polyethylen auf einem Freifallscheider unter besonderer Berücksichtigung chemischer Konditionierung“. Dissertation, Shaker Verlag, Aachen, 2000

[Sch02] Schubert, G.: „Zerkleinerungstechnik für nicht-spröde Abfälle und Schrotte“. In: Aufbereitungstechnik, Vol. 43 (9), pp. 6–23, 2002

[Sch11] Schöppner, V.: „Grundlagen der Kunststoffverarbeitung“. Vorlesungsskript, Universität Paderborn, Kunststofftechnik Paderborn (KTP), 2011

[SNS+04] Schubert, G.; Németh, E.; Simon, F.; Albrecht, V.: „Beeinflussung der Kontaktaufladung von Kunststoffen durch Modifizierung ihrer Oberfläche“. In: Partikelwechselwirkung bei Prozessen der Mechanischen Verfahrenstechnik, Husemann, K.; Graichen, K. (Herausgeber), Shaker-Verlag, Reihe: Berichte aus der Verfahrenstechnik, 183, Aachen, 2004

[Sti97] Stieß, M.: „Mechanische Verfahrenstechnik 2“. Springer-Verlag, Berlin, Heidelberg, 1997

[Sti09] Stieß, M.: „Mechanische Verfahrenstechnik – Partikeltechnologie 1". Springer-Verlag, Berlin, Heidelberg, 2009

[THB+22] Tölle, L.; Hopp, M.; Bünger, J.; Westphal, G.; Rosenkranz, N.; Monsé, C.: „Faserstäube beim FVK-Recycling". In: Kunststoffe, Vol. 5, Carl Hanser Verlag, München, 2022

[TH21] Tölle, L.; Hopp, M.: „Influence of Fiber Content on the Fiber Dust Formation During Shredding for Mechanical Recycling of Different Glass and Carbon Fiber Reinforced Thermoplastics". 36th International Conference of Polymer Processing Society (PPS), 2021

[TH22a] Tölle, L.; Hopp, M.: „Influence of process parameters on the formation of inhalable fiber dust during shredding for mechanical recycling of fiber-reinforced organo sheets". In: Journal of Applied Polymer Science, Vol. 139 (29), e52646, DOI:10.1002/app.52646, 2022

[TH22b] Tölle, L.; Hopp, M.: „Effect of shredding parameters on the formation of airborne fiber dust during mechanical recycling of fiber reinforced thermoplastics". 20th European Conference on Composite Materials, Lausanne, Switzerland, 2022

[THH21] Tölle, L.; Hopp, M.; Hoemann, T.: „Reduzierung der Faserstaubentstehung beim mechanischen Recycling – Zerkleinern von carbonfaserverstärkten Polycarbonaten mit unterschiedlichen Verstärkungsformen". In: GAK Gummi, Fasern, Kunststoffe, Dr. Gupta Verlag, Vol. 11, pp. 478–483, Ratingen, 2021

[TMR+22] Tölle, L.; Monsé, C.; Rosenkranz, N.; Haibel, N.; Walter, D.; Bünger, J.; Hopp, M.; Westphal, G.: „Characterization of fiber dust resulting from recycling of Carbon fiber-reinforced thermoplastics (CFRP) and their cell toxicity". In: Journal of Materials Science and Chemical Engineering, Vol. 10, pp. 1–16 2022

[TPK+19] Tschachtli, S.; Pitschke, T.; Kreibe, S.; Martin, A.: „Recyclingfähigkeit von Verpackungen – Konkretisierung Untersuchungsrahmen und Kriterienkatalog". Bericht im Auftrag der Interseroh Dienstleistungs-GmbH, Augsburg, März 2019

[UB97] Uhlig, D.; Bremerstein, I.: „Handbuch der Sortiertechnik". Hrsg.: Duales System Deutschland, Verlag TÜV Rheinland, 1997

[Wol04] Woldt, D.: „Zerkleinerung nicht-spröder Stoffe in Rotorscheren und -reißern". Dissertation, Technische Universität Bergakademie Freiberg, 2004

4 Material Recycling

In addition to feedstock recycling, material recycling is classified under the recycling category according to the waste hierarchy of the EU Waste Framework Directive. At just over 46 % (based on the previous calculation point (1), see 2.4), it is, alongside energy recovery, the most important recovery route for plastic waste in Germany. In addition, there are by-products amounting to 639,000 tonnes, which also undergo material processing. The average increase in material recycling relative to the respective prior-year values amounts to 3.6 % over the period from 1994 to 2021 [Con21, p. 106].

After a brief overview, the following section will first address the recycling product most apparent to the consumer, the PET bottle and its recycling route. In order to increase the share of material recycling in the future, it is necessary to design products for recyclability already during the product development process, which is the subject of Section 4.3. In 1.3.1, the fundamental prerequisites for obtaining recyclate were already discussed, and it was also briefly mentioned that specific requirements exist for the material to be used in various application areas. At this point, Section 4.4 will address two specific industries and the resulting requirements for recyclates. Recyclable products, at their end-of-life stage, ideally lead to a waste stream whose processing is economically and technically feasible. In addition, there is the possibility that the recyclates produced from it also meet the requirements for industries that have hitherto been difficult to supply.

Some aspects of processing recyclate will be addressed subsequently. This is followed by explanations on the separation of composites from multi-component parts, which on the one hand concern the current state of research and on the other hand present original, fundamental investigations into the separation of material-bonded composites.

4.1 Overview and Definitions

In material recycling, plastics are extracted from the waste stream in such a way that their polymer structure is preserved and, after processing, is available as recyclate for the manufacture of new products. This is also the origin of the term material recycling: the plastic remains in the cycle as a material and is reused after processing.

The processes for material recycling can in principle be divided into mechanical and solvent-based processes. Mechanical recycling, which is also frequently used as a synonym for material recycling, encompasses the classical processing steps described in Chapter 3, including sorting, size reduction, cleaning, separation by plastic type, drying, and re-granulation. Solvent-based processes operate on the principle of dissolving the plastics. In this process, solvents suitable for the respective plastic types penetrate the polymer structure and reduce the secondary valence forces between the polymer chains. No chemical reaction takes place and the polymer chains retain their structure. Through filtration steps and removal of the solvents, the polymeric composite is ultimately reconstituted through the action of secondary valence bonds. These processes are used for the separation of material-bonded composites (selective dissolution), which is discussed in more detail in Section 4.6.5, or for the removal of additives from the polymeric composite. Even though the polymer chains in principle retain their structure in material recycling processes, unintended material degradation, i.e. shortening of the polymer chains, occurs during the processes. This is attributable in particular to thermo-mechanical, thermo-oxidative, and, in the case of polycondensates such as PET, hydrolytic influencing factors.

Figure 4.1 schematically shows the stages of the plastics cycle as it presents itself in material recycling.

In the production phase, products are developed and manufactured that are ideally recyclable. Where possible, this is done using a high proportion of recyclate, to which a portion of virgin material may be added if necessary.

The products are sold and, upon purchase by a consumer, enter the use phase. Ideally, the product is reused multiple times to avoid waste and ultimately enters disposal after use. This is followed by the processing phase, which in this case is characterized by material recycling. During the sorting, washing, and separation processes, material losses occur that are directed to energy recovery.

Overall, it can be said that the following points along the cycle will play key roles in the coming years in order to further expand material recycling in line with the politically stated targets:

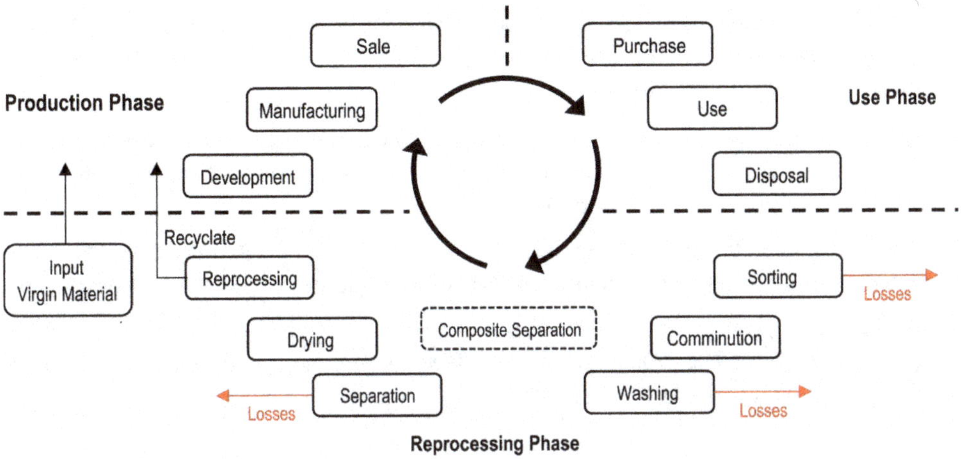

Figure 4.1 Overview of the circular economy processes in material recycling

Development in the production phase: Design for Recycling will gain significantly in importance so that more plastics can be recovered from waste. This applies not only to packaging but also to technical components.

Disposal at the end of the use phase: Separation by the consumer determines the routes that the waste takes in sorting and processing facilities. In the future, consumers will again be held more accountable in this regard. In addition, further collection routes will need to be considered, for example through the expansion of deposit systems or through other forms of take-back systems, including for the industrial sector. The logistics along the entire disposal pathway up to the sorting facility or recycler are decisive for the quality of the waste stream to be processed.

Composite separation in the processing phase: By separating material composites, significantly more plastics can be extracted from the waste stream and high-quality recyclates can be produced. In recent decades, the trend has been to implement multi-material mixes in plastic components as functional integration. This should now be made recyclable by developing, improving, and economically implementing the corresponding processes (see the explanations in Section 4.6.5).

In the following, a few terms are briefly explained that are frequently used in German-language usage for a material-processed plastic material, according to [HHB10, p. 55].

- Recyclate: generic term for processed plastic material with defined properties that has already undergone a processing step; recyclate can be regrind, agglomerate or compactate, regranulate, and regenerate.
- Regranulate: plastic material that has been completely plasticized and homogenized in the processing procedure, in the form of granulate, without the addition of further additives

- Regenerate: Plastic material that has been completely plasticized and homogenized in the processing procedure, upgraded with additives in a compounding process, and is available in granulate form
- Agglomerate: Partially plasticized, conditionally homogenized plastic material that has already undergone a processing step and frequently enters the agglomeration process as a mixed plastics fraction
- Regrind: Recyclate produced by intensive size reduction, i.e., grinding. The particle size distribution is broadly scattered; regrind frequently contains a proportion of dust.

Frequently, however, the word recyclate is used in general for all possible forms of processed material without further distinction.

4.2 Best Practice: Bottle-to-Bottle

For many years, the PET bottle has been regarded as a model example of the principle of a true product closed-loop system with an organized collection and a high return rate. Of all food packaging materials, the PET recyclate obtained from PET bottles under strict conditions is so far the only material that may be reintroduced into direct food contact as a PET bottle. Even though PET recyclate is still used at an unexpectedly low share of 44.7% for bottle production, partly for cost reasons, the potential exists to increase this share to 93–97% with regard to technical properties. It is therefore worthwhile to take a closer look at this particular material stream, which will be presented in the following [NN07-ol].

In 2019, 467,400 tonnes of PET beverage bottles were consumed in Germany, corresponding to a quantity of 18 billion PET bottles. Approximately 87% of this quantity consists of single-use bottles subject to a deposit obligation, whose return rate through the deposit system is very high at 98%. This demonstrates that the deposit redemption through organized return provides a strong incentive for consumers and that this regulatory instrument functions well. Non-deposit PET beverage bottles enter the recyclables collection system to a predominant extent and thus into the dual systems. Sorting losses of approximately 9,400 tonnes remain, which, in addition to losses from processing, are directed to energy recovery. The PET recycling quantity produced in 2021 from PET beverage bottles amounted to approximately 446,000 tonnes, of which 44.7% were processed into bottles, 26.8% into films, and 11.3% into fibers for the textile industry. The remaining share is attributable to PET products from various applications. With regard to the new production of PET bottles, the recyclate content in 2021 was 44.8% (figures from [NN07-ol]).

Figure 4.2 shows an overview of the individual process steps in the processing of PET bottles from the deposit return system. The gray-shaded fields show the components of the waste stream separated in each process step. After bale opening, metallic components are separated. Ballistic separators ensure the separation of loose labels or films. A subsequent first color sorting removes the colored PET bottles. After the following size reduction, the PET flakes are freed from labels, paper, and adhesives in a hot wash. The float-sink separation then allows the screw caps made of PP or HDPE to be separated from the PET stream. Drying and air classification follow to free the PET flakes from surface-adhering contaminants for the URRC process.

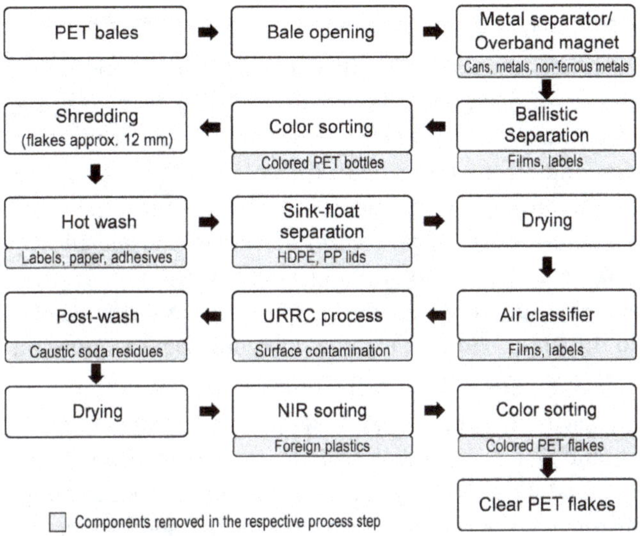

Figure 4.2 Overview of the process steps in the processing of PET for renewed use in bottles with food contact

URRC stands for "United Resource Recovery Cooperation" and is a patented process developed to market maturity with the Coca-Cola Company for processing plastics with hot caustic soda. This process qualifies the PET flakes for reuse in packaging with direct food contact and is thus one of the processes required to obtain approval of a recycling process from the EFSA (European Food Safety Authority). In the process, the flakes are wetted with dilute caustic soda and dried. Surface cleaning then takes place in a rotary kiln at various temperature stages, during which the outermost polymer layer is cleaved off. In the subsequent post-wash, the material is neutralized by washing off the caustic soda residues. After a further drying step, the flakes are first sorted by plastic type and then the pure PET fraction is sorted by color [NN09-ol].

The highly pure input stream arriving at the recycler through the deposit system enables efficient recycling. The URRC process serves to upgrade the material for renewed use as food packaging. Due to the higher prices for virgin PET compared to LDPE, the comparatively expensive processing procedure for PET is economically worthwhile.

4.3 Design for Recycling

In order to ensure the long-term availability of high-quality recyclates in sufficient quantities, influencing the input waste stream with regard to a recyclable design of products is indispensable. Only if plastic products meet certain requirements in terms of their material composition, external shape, labeling, and other factors is material recycling possible with respect to the sorting and processing plants currently in operation. The following section addresses the requirements for recyclability, the resulting design guidelines with regard to the packaging sector, and the necessary certification.

In general, design for recycling should be integrated into every product development process. However, to date the focus of discussion has been primarily on packaging, because on the one hand it represents the largest share by volume in the production and waste generation of plastics, and on the other hand it has long been the subject of public debate in connection with environmental pollution. For this reason, design guidelines that enable material recycling are in many cases discussed and specifically further developed only for packaging. In order to also address EU legislation and the prioritization of the waste hierarchy, guidelines for design for reuse and further use strictly speaking also belong to this topic. However, since this book focuses more on actual recycling than on reuse, elaborations on this are omitted.

Although the main focus in the discussion of recyclable design lies in the packaging sector, at least two guidelines should be mentioned that address the topic of design for recycling in general. In 2002, VDI guideline 2243 "Recycling-oriented product development" was published, providing developers and designers with guidelines for increasing the recyclability of technical products. Endowed with legislative significance is the Ecodesign Regulation (Regulation 2024/1781 of the European Parliament and of the Council of 13 June 2024 establishing a framework for setting ecodesign requirements for sustainable products). It takes up the approaches from the Ecodesign Directive and establishes minimum criteria for a broad range of products regarding

- Energy and resource efficiency,
- Recyclate content,
- Ease of disassembly, repairability, durability,
- Reusability and recyclability.

In addition, all products falling under the regulation are to have a digital product passport that provides consumers with information on circularity and sustainability ($CO_2$and environmental footprint).

4.3.1 Recyclability

Under the Packaging Act, the dual systems are obliged, in the context of setting participation fees, to create incentives for the production of packaging that can be recycled to a large extent. The minimum standards for assessing recyclability are established by the Stiftung Zentrale Stelle Verpackungsregister, headquartered in Osnabrück. It defines the term recyclability as follows: "[This] recyclability is the fundamental and gradual suitability of packaging, after passing through industrially available recovery processes, to substitute virgin material in material-typical applications" [NN08-ol, p. 7]. The issue is therefore fundamentally, firstly, the nature of the packaging as such and, secondly, the actual recovery pathway after the use phase for the specific packaging. Further recognized specifications regarding material recyclability are contained in the standard DIN EN 13430: "Requirements for packaging recoverable by material recycling".

According to the minimum standards of the Zentrale Stelle Verpackungsregister, packaging must in principle be assessed for

- Whether a suitable sorting and recovery infrastructure is available,
- How well it can be sorted and separated, and
- Whether recycling incompatibilities (e.g., PET sleeves on HDPE bottles) are present.

The systems may add further criteria to assess recyclability.

To be distinguished from the concept of recyclability is the life cycle assessment. This encompasses the assessment of all processes from production to recovery, whereas recyclability focuses only on the post-use phase. When comparing two similar products, the life cycle assessment of one product may be more favorable than that of the other product, while the first may exhibit poorer recyclability. Recyclability thus constitutes an independent assessment parameter.

4.3.2 Design Guidelines for Packaging

In the following, some design guidelines will be presented by way of example that lead to the recyclability of packaging. In addition, features that prevent recycling are also shown. The following compilation makes no claim to completeness, and the obstacles to recycling may also be quickly superseded by further developments in the state of the art in sorting and processing technology. Moreover, it is frequently details that determine recyclability, so only key features can be mentioned here by way of example.

Multilayer/Monolayer

In principle, packaging should consist of only one plastic material in order to generate a single-variety waste stream. As explained in Section 4.6.2, material-bonded plastic/-plastic composites present major challenges in terms of the separability of the components. In some cases, separation can be achieved by disintegration size reduction; however, this does not apply to the typical multilayer films frequently used in food packaging to extend shelf life. In addition to processing technology, sorting technology plays an essential role in the first instance. Multilayerfilms currently cannot be adequately detected in their compositions by NIR spectroscopy, so they are sorted into the mixed plastics fraction. A monolayer packaging should therefore be aimed for wherever possible. Barrier effects can also be achieved via SiO_xplasma coatings [PSW+11] or can be produced by biaxially oriented films.

An exception for a tolerable multilayer structure exists for PP/PP-EVOH. According to studies by RecyClass, EVOH has only a minor impact on reuse if the EVOH content is less than 5% [NN10-ol].

Plastic Selection with an Established Recycling System

In currently operating light packaging sorting facilities, waste pieces are separated into the pure plastic fractions PP, PE, PS, PET or into the mixed plastics fraction. The latter is not available for high-quality recycling. Therefore, packaging should consist of one of the mentioned plastic types. Proportions of PVC, POM, PA or other plastics must therefore not be contained in the packaging or must be classifiable as a subordinate, separable contaminant.

Modification of Plastics

Plastics are modified in various ways for the processing phase and for the use phase in order to adapt the properties to the application. Here too, it must be noted that the modification can prevent sorting into the fractions or the subsequent processing.

Blowing agents can, for example, cause plastics to no longer be separable in a float-sink separation, since the density of the plastic is artificially reduced by the foaming of the packaging. Density-increasing fillers lead to the same challenges regarding separability. Fillers should therefore only be used in low concentrations.

Colorations that cause complete absorption of light, as is the case with the use of carbon black, are to be avoided. These cannot otherwise be detected by NIR spectroscopy. Black packaging can now be realized with so-called NIR-detectable black masterbatches. In general, light colors are to be preferred.

Labels and Sleeves

If labels are not made of the same material as the packaging itself, they must not cover more than half of the surface. For PET bottles, it further applies that the density of the label material must not be greater than 1 g/cm^3, so that density separation using water as the separating medium functions correctly.

Printing Inks and Adhesives

Printing inks must not contain any hazardous substances listed on the exclusion list of the European Printing Ink Association (EuPIA). Adhesives for labels must be water-soluble or, in the case of PET, at least soluble in an alkaline environment at 80 °C.

4.3.3 Certification of Recyclability

On the basis of the minimum standards already mentioned by the Zentrale Stelle Verpackungsregister and on the basis of DIN standards for requirements on material recycling and corresponding labeling, packaging can be objectively evaluated. The fundamental criteria and steps of the evaluation are:

- Assignment to accepted material according to fraction no.
- Recyclability in established collection and recycling structures
- Existence of sorting infrastructure
- Identifiability via sensor-based sorting
- Separability
- Recycling incompatibilities

For the individual steps, the various providers have established concrete testing schemes according to which a detailed evaluation can be carried out. In many cases, a percentage value is indicated for classifying the recyclability of the packaging. As examples, Figure 4.3 some certificates are shown that can be requested by those placing the packaging on the market.

Figure 4.3 Certificates for declaring recyclability

4.4 Requirements for Recyclates

The market for recyclates has been considered difficult for many years, as virgin material is in many cases cheaper than the recyclate. In addition, there are a wide variety of requirements from different applications that recyclate manufacturers must also meet. In 1.3.3, the prerequisites that must be in place to obtain a desired plastic as a recyclate from waste are derived from the perspective of a product developer. In contrast, the following discussion concerns specific requirements placed on recyclates.

Important for the processing of high-quality recyclates is the availability of a pure-grade input stream and a coordinated processing method. For the first point, design for recycling is essential, since only through recyclable products can the material even be available to produce high-quality recyclates. The technologies for processing are available for numerous applications; however, economic viability plays a significant role in processing quality due to competition with virgin material.

The requirements for recyclates are diverse and can, among other things, be specified

- By the state through laws and regulations,
- By industry associations,
- By producers or
- By machine manufacturers.

A distinction must be made as to whether these are purely technical requirements such as strength, elastic modulus, MFI value or chemical resistance, or specific requirements designated especially for recyclates such as the presence of foreign substances in the ppm range or proof of health safety. The general specification "Use of virgin material only" in specifications from producers to their suppliers is becoming less common. Rather, it is required that the technical requirements must be met as with virgin material. In addition, higher standards with regard to quality management are required. In the following, the specifications arising from the REACH regulation will first be examined, and then the requirements from two industrial fields will be briefly outlined as examples.

4.4.1 REACH and Recyclates

REACH stands for Registration, Evaluation, Authorisation and Restriction of Chemicals and represents a European regulation (1907/2006) that came into force in 2007 and has since been subject to numerous amendments. According to this chemicals regulation, manufacturers, importers and downstream users of chemical substances are obliged to carry out a safety assessment for the substances and to pass the relevant information along the supply chain. This concerns safety for human health and the environment. In addition, those placing the substances on the market must have them registered or demonstrate that they are already registered. These regulations apply when the relevant substance quantity is at least one tonne per year.

Products from the plastics industry also fall under the REACH regulation, and thus so do recyclates. Waste, on the other hand, is not regulated under the REACH regulation. For the plastics industry, it therefore generally applies that placing virgin material on the market falls within the responsibility of the manufacturer. If the plastic temporarily becomes waste, the regulations of the waste ordinances and laws apply, whereby the safety data sheet to be issued by the manufacturer must also contain information on disposal. If the plastic material leaves waste status again as a recyclate, the responsibility for registration and the issuance of a safety data sheet lies in principle with the respective recycler and no longer with the manufacturer of the original virgin material. When waste again becomes a substance within the meaning of the REACH regulation is not always straightforward to answer. The legal concepts of the Waste Framework Directive and the REACH regulation intersect at this point. Under REACH, waste only becomes a substance subject to the REACH regulation again when ownership is transferred from the recycler to the purchaser [GKH21, p. 14] or when "final recovery" takes place [JRS+11]. A case-by-case distinction depending on whether it is a substance, a mixture or an article under REACH is necessary. The so-called recycling privilege, which exempts the recycler from the registration obligation under certain circumstances, must also be taken into account in this context. According to Article 2(7)(d) of the REACH regulation, the conditions for exemption are that the recovered substance is identical to an already registered substance and that the relevant hazard information from the safety data sheet is available.

In the first step, the recycler must ascertain whether the recyclate is identical to an already registered substance. He does not need to pay attention to who manufactured the starting substance, since the supply chain was in any case interrupted when the original substance became waste. Substances are considered identical under REACH if the main constituent, for example propylene as the monomer of PP, is contained in the substance at 80%. The recyclate may therefore contain additives and impurities up to a proportion of 20%. However, if these are intended to fulfill a defined function in the substance or mixture, the registration obligation also applies in principle here, unless Article 2(7)(d) also applies in this case. If the recycling privilege applies, substances do not need to be registered and no extended safety data sheet needs to be prepared. However, it is still advisable to pass on handling instructions to purchasers. In addition, all information must be available at the recycler that justifies the application of Article 2(7)(d). If, however, so-called substances of very high concern (SVHC) are contained in the recovered substance, authorization must be obtained here if it has not already been granted [GKH21].

If the recycling privilege does not apply, the contained substances must be analyzed and checked as to whether they are already registered as a substance. If this is not the case, registration must be carried out and a safety data sheet must be prepared with an assessment of the hazard potential of the foreign substances and impurities.

The information obligations under REACH require the establishment of a suitable QA system, which does not necessarily have to conform to a standardized standard. According to [JRS+11] should, however, contain the following elements:

- Incoming inspection of the waste input stream
- Visual inspection, analytics, acceptance test certificates, retained samples
- In-plant traceability of the material

  •→ appropriate labeling, separate storage, deliberate consolidation with coupling of information
- Process control

  •→ process monitoring throughout the entire processing operation
- Outgoing inspection

  •→ testing of in-plant defined specifications, retained samples
- Complete documentation

  •→ assignability of measurement data, records, process data

The application of the REACH regulation for recycling companies requires a more in-depth engagement with the subject. Only the basic principles could be addressed here, as case-by-case considerations are necessary. Guidance for recyclers can be found, for example, in [JRS+11] and [GKH21].

4.4.2 Recyclates for Food Contact

For direct contact with food, there are strict requirements regulated by the European Food Safety Authority (EFSA). The German Federal Office of Consumer Protection and Food Safety (BVL) serves as the interface to the EFSA for plastics and recyclate manufacturers. In principle, no packaging or finished products are certified; instead, the basic substances contained in the product are tested and approved. For the recycling sector, the "Regulation (EC) No. 282/2008 on recycled plastic materials and articles intended to come into contact with foods [...]" governs the specific provisions for this application.

The following rules apply to the operation of a recycling process for recyclate intended to come into contact with food:

- The recycling process must be authorized.
- The authorization holder must implement a suitable QA system.
- The authorization holder must notify the respective production sites to the EFSA via the BVL.

The basic substances contained in the recycled plastic must be listed in the so-called "Union list" (Annex 1 of the above-mentioned regulation). If the substance is not listed there, an authorization procedure must be initiated, on which the European Commission decides. The recyclate must furthermore have been freed from contaminations to such an extent that no risk to human health exists, which must be demonstrated by analyses. Alternatively, the material originates from a production of products already intended for food contact, in which any contamination can be excluded through a monitored process. In addition, the criteria for quality monitoring and the conditions of use must be established so that no constituents can be transferred to food.

4.4.3 Recyclates for the Electronics Industry

In the electronics industry, special requirements such as dielectric strength, tracking resistance, flammability and fire behavior, or the ignitability of plastics play a major role. The properties specifically adjusted via additives must be maintained even under critical operating conditions (temperature, climate). Material purity is decisive for this, as defects tend to promote increased migration of additives. This in turn can locally lead to lower proportions of functional additives and to failure of components, particularly in the event of electrical breakdown.

There are numerous test standards for applications in the electronics industry which, although not mandatory in the EU, are a prerequisite for the import of products particularly in the USA. UL certification is frequently required in this context. The following standards are particularly important for plastics:

- UL 746 A – Polymeric Materials – Short Term Property Evaluations
- UL 746 B – Polymeric Materials – Long Term Property Evaluations
- UL 746 C – Polymeric Materials – Use in Electrical Equipment Evaluations
- UL 746 D – Polymeric Materials – Fabricated Parts
- UL 94 Tests for flammability of Plastic Materials for Parts in devices and Appliances

For recyclates, UL 746 D specifies that for recyclates with complete material traceability, all tests in accordance with UL 94 and UL 746 A must be carried out on three batches. If the origin of the material cannot be traced, the tests must be carried out on five batches. In addition, there is an obligation to establish a suitable QA system.

4.5 Processing of Recyclate

In principle, recyclates can be processed using the same processing machinery and the process conditions customary in plastics technology. No fundamentally different machines are required to process recyclate into pipes, films, or injection-molded components. For material pre-treatment and feeding, additional units may need to be connected upstream, depending on the condition in which the recyclate is to be processed.

Nevertheless, challenges arise during processing. Some of these will be discussed in the following.

4.5.1 Challenges in the Processing of Recyclate

As already described in this book, depending on the material origin and the available collection, sorting, and processing technology, recyclates are produced with varying quality levels. PET recyclate processed for bottle-to-bottle recycling differs only marginally from the corresponding virgin material, whereas the processing of recyclate from the mixed plastics fraction of a light packaging sorting plant with a manageable number of processing steps leads to a number of challenges during processing.

The quality of the processing is in principle dependent on the price structure in the market and on the technical requirements that potential customers specify to the recycler. If the waste is insufficiently processed, fluctuations in the composition of the material can occur. For example, strongly adhering contaminants on parts of the batch can influence the flowability of the material. Insufficient removal of foreign plastics also directly affects the viscosity and melt density of the material. If these influences are not constant over time within a batch, fluctuations occur during the processing operation. If these are so severe that the specification is no longer met for certain characteristics, this leads to rejects. If good melt homogenization has been achieved during recyclate processing, a consistent process within a batch may still be possible despite the contaminations. In this case, it may only be necessary to adjust the process parameters from batch to batch.

To respond to process fluctuations, two basic strategies can be distinguished:

- Use of robust plant technology and achievement of a moderate throughput

  •In injection molding, buffer times during the residual cooling time can be used to extend the metering process and subsequent thermal equalization processes in terms of time, thereby ensuring in every case a thermally and materially homogeneous melt. In extrusion, when selecting the extruder or designing the screw, attention should be paid to reserves in plasticizing capacity. Melt filters and degassing points ensure greater process consistency.

- Sophisticated inline measurement technology and real-time control

  Another approach is the elaborate acquisition and monitoring of relevant process variables such as pressure and temperature, or the integration of further target variables such as the automated recording of component weight or geometric deviation. Using more recent methods, predictions can be developed, for example from the flow behavior of the melt, via machine learning methods, by using a large number of process and machine data as a so-called soft sensor. In this way, cause-and-effect relationships between characteristic variables and target variables can be identified. For example, a material property predicted during plasticization in injection molding can bring about a corresponding real-time control of the injection speed and lead to the maintenance of the desired quality [MDK+20].

With the first strategy, recyclates can often be processed without major effects on the manufacturing-related quality of the parts. The second strategy is complex and cost-intensive. In addition, it must be determined whether the detection of the fluctuations can be measured with sufficient accuracy so that control can be implemented as promptly as possible.

4.5.2 Melt Filtration

Melt filters are an independent component of an extrusion system and serve to remove contaminants from the plastic melt. During the extrusion of virgin material, these can be, for example, impurities from the periphery of the granulate supply due to wear or from the environment. During recyclate processing, contaminants can additionally be contained in the granulate that cannot be completely separated during processing. They therefore play an essential role in systems on which recyclate is primarily processed.

The filters are mostly installed between the extruder and the extrusion die in corresponding housings and connected to the adjacent system components by flanges. When designing the filters, the required fineness of the filter medium and the filter area are relevant quantities to be determined. High levels of contamination lead to a short service life of the filter medium at small filter areas and high filter fineness, which then has to be replaced. The condition of the filter is monitored via the differential pressure upstream and downstream of the filter using corresponding pressure sensors installed in the filter line. In Figure 4.4 On the left, the schematic structure of the pressure measurement in the system is shown. If the differential pressure exceeds a certain threshold value, the filter insert must be replaced.

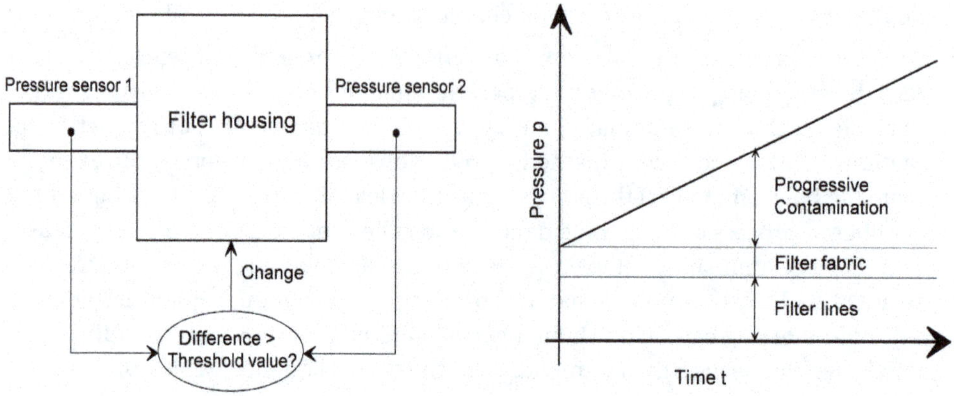

Figure 4.4 Structure of the pressure measurement (left) and pressure profile with progressive filter contamination (right) [Sch20, p. 75]

In the diagram on the right, the total pressure loss in the filter can be seen qualitatively. It is composed of the two constant pressure components, caused by the filter lines and the uncontaminated filter medium, and the pressure component that increases linearly with time, which is caused by the contamination [Sch20].

There are various designs of filters, the differences of which essentially lie in the extent to which production is disrupted by changing the filter medium. For example, in the simplest design, in which a filter medium is placed in front of a permanently installed perforated plate, production must be completely interrupted when the filter is to be changed (see Figure 4.5, left).

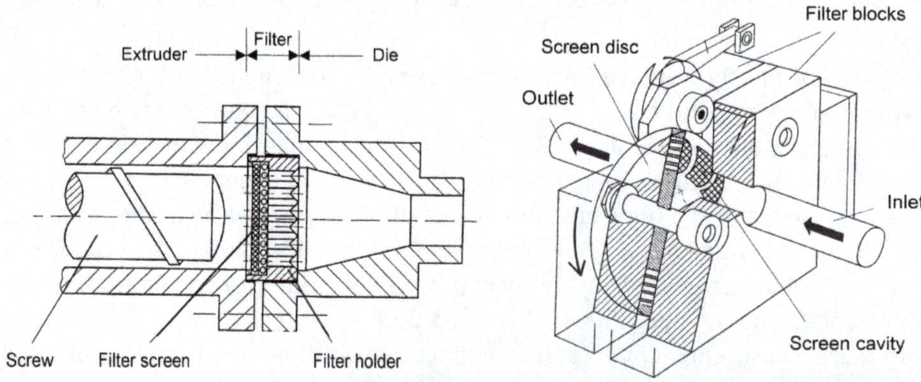

Figure 4.5 Simple, permanently installed filter in front of a perforated plate (left) [BMW97, p. 109] and screen wheel filter by Gneuß (right) [BMW97, p. 110]

This is only useful with frequent product changes or with large-area filters whose service life is designed to be correspondingly long. In addition, there are continuous filters that allow the filter medium to be changed during production operation. The designs range from so-called screen wheel or rotary filters (see Figure 4.5, right) via piston screen changers to filter drums. Backflush systems or continuous scrapers ensure pressure-constant operation in complex systems. The former ensure backflushing of one of the screens by redirecting a portion of the already filtered melt stream, whereby the contaminants can be removed from the screen. This increases the service life of the filters and the intervals for filter changes, thereby reducing screen costs. In other systems, rotating scrapers in front of the filter continuously ensure the removal of contaminants. With simpler designs, production is not interrupted, but the pressure drop leads to rejects during the filter change [NN11-ol].

A filter used in the recycling sector at high contamination levels is in Figure 4.6 schematically using the example of type BKG® HiConTM R-Type 250 from the manufacturer Nordson BKG GmbH. The contaminated melt stream enters the gap between the cylindrical separating head and the circumferentially arranged filter basket. The knives arranged at an angle on the separating head scrape the contaminants off the filter and convey them to the downstream discharge screw. The melt heavily contaminated with impurities is in this way removed from the process, while the clean melt passes through the filter and enters the shaping die. According to company information, throughputs between 500 and 1500 kg/h and filtration fineness down to 80 µm can be achieved at a maximum pressure of 350 bar [NN12-ol].

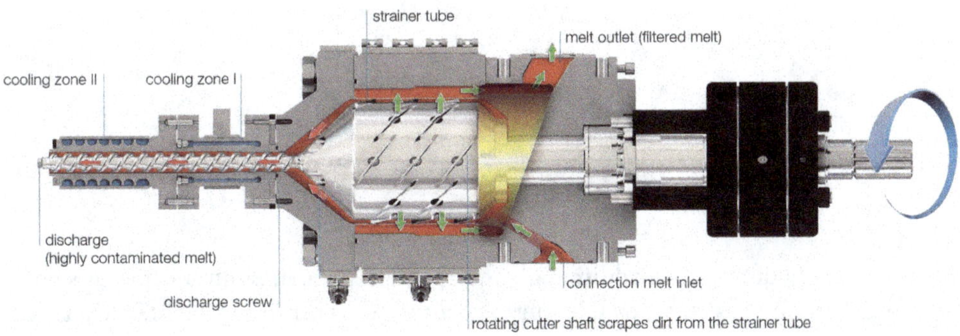

Figure 4.6 Functional principle of the drum filter BKG® HiConTM R-Type 250 by the manufacturer Nordson BKG GmbH [NN12-ol]

4.5.3 Odor Absorption

In order to produce high-quality recyclate, the removal of disturbing odors is an important prerequisite in some cases. Substances remaining in the plastic that cannot be removed by simple washing processes and that lead to disturbing odors result from diffusion processes of low- or high-molecular-weight substances into the plastic. The substances originate from food, cosmetics, or cleaning agents. However, decomposition products, e.g. from oxidation or hydrolysis processes, can also lead to undesirable odors.

Slightly volatile low-molecular-weight substances can be removed from the melt via a degassing process in a co-rotating twin-screw extruder. Using a two-stage screw concept, after the plastic has melted, a decompression is established through a strongly increasing flight depth of the screws, which allows a high exchange surface between the ambient air and the melt. In addition, the ambient pressure prevailing there prevents the melt from escaping (see Figure 4.7). Low-molecular-weight substances can in this way escape from the melt and transition into the gas phase. In addition to atmospheric degassing, a higher degassing performance can be achieved with vacuum degassing. Recycling manufacturers select different technologies and process steps depending on the input stream in order to achieve effective odour reduction.

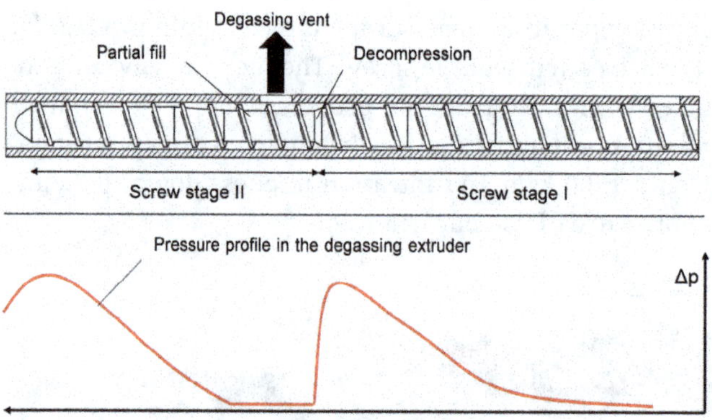

Figure 4.7 Structure of a degassing screw with associated pressure profile [Sch20, p. 60]

The removal of odors can be additionally improved in the compounding process with so-called entraining agents. For this purpose, a suitable entraining agent, such as e.g. supercritical CO_2, added to the melt under high pressure. The CO_2 has good solubility in the melt and absorbs the volatile components. In the vacuum degassing process currently in use, the stripping agent is discharged together with the volatile, odour-causing substances [JD14].

Odors can also be removed from bulk material in a material-friendly manner by first heating it to a temperature of approximately 60 °C or higher, depending on the plastic, and then flushing it with hot gas in a silo. The higher temperature allows diffusion processes to proceed more quickly, and the hot gas absorbs the volatile components. The process can take several hours until the desired result is achieved. Therefore, the silos must be designed to be correspondingly large or set up redundantly [NN13-ol; SS22-ol].

4.6 Composite Separation of Multi-Component Products

In the simplest case, the manufacture of plastic products uses only one type of plastic, whereby a single-component, materially homogeneous and, in the end-of-life state, unmixed component is fed into the use and processing phase. The material used may contain fillers or reinforcing agents. However, most components must exhibit different properties at different locations, which must be fulfilled by the selected plastic. These can be mechanical, optical or haptic properties, or those relating to heat resistance or chemical resistance. At locations of the component with high mechanical requirements, high-performance plastics may therefore be necessary, while this is not the case at other locations. The combination of several types of plastic offers potential savings. Further characteristics that affect only a small part of the component, namely the surface, are decorative quality, haptics and friction behaviour. The properties of the different types of plastic can be combined by parallel or series connection, so that a component adapted to the respective application is created in multi-material design. In addition to plastics, metals, ceramics, textiles or paper can also be used as composite partners.

Shown as examples in Figure 4.8 are multi-component products in which different requirements have led to the necessity of a multi-component structure. In the application as a multilayer pipe foamed middle layers are produced by co-extrusion as sound insulation and for weight reduction in drainage pipes, EVOH layers as oxygen barriers or PA layers as hydrogen barriers for gas-carrying pipes, or reinforced PP outer layers for use in the ground. The second example shows an overmolded particle foam that functions as a damping element. The combination of a housing with a directly injection-molded seal has meanwhile found widespread application, as can be seen exemplarily in the image at the top right. Particularly in the automotive industry, not least due to electric mobility, there is a necessity to pursue structural lightweight construction. For this purpose, fiber-reinforced plastics are frequently used, specifically fabric-reinforced so-called organo sheets, which exhibit very good weight-specific mechanical properties. For the combination of these with other components, joining technologies are necessary, such as the rivet connection shown in the image at the bottom left. The image at the bottom center shows an assembly produced in a special injection molding process, consisting of a cut, preheated and formed organo sheet reshaped in the injection molding tool and an injection-molded media line. This is inflated to a defined inner diameter during the injection molding process by two sequentially connected gas injections. The process used for this is described in the literature as the GIT-Blow process and was developed at the Kunststofftechnik Paderborn at the University of Paderborn. The combination of a rigidly designed carrier element in conjunction with a media line is an example of functional integration and the consolidation of manufacturing steps into an overall process. The last product example concerns the widely discussed multilayer film, which is used in the field of food packaging to ensure shelf life. It consists of a combination of several plastics, each fulfilling different functions such as gas or water vapor barrier or printability.

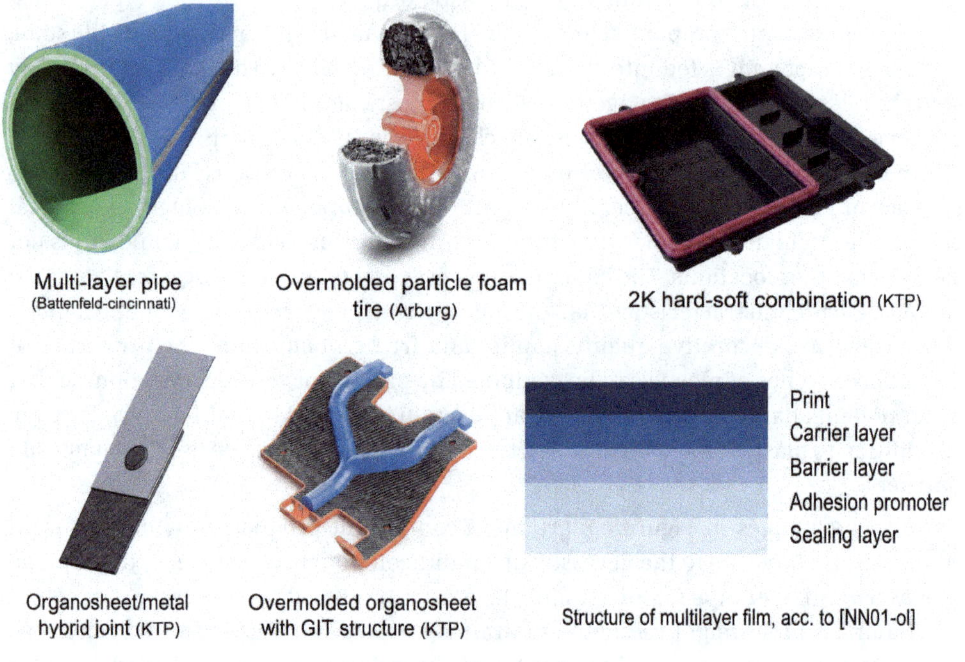

Figure 4.8 Examples of multi-component products

From the perspective of plastics engineers, the possibilities of multi-component technology and the associated innovations that have emerged over recent decades form the basis for producing complex hybrid assemblies or intelligently structured composite structures. However, these engineering achievements run contrary to the requirement of providing pure material streams at the end of the life cycle for material recycling. In particular, bonded composites are fed into sorting facilities as mixed plastics fractions, which are mostly subjected to energy recovery. The separation process of these mixed fractions is complex and has so far only been possible within certain limits. In the following, the types of composites and their challenges in recycling are presented. In addition, with regard to bonded composites, current research approaches are presented that aim at the composite separation of multi-component parts.

4.6.1 Composite Types

In principle, three categories of composites can be distinguished with regard to the mechanism of cohesion [Sch22]:

Form-fit connections achieve their cohesion through an undercut in the direction of force that prevents the separation of the two bonding partners. Examples of form-fit joining methods are nailing, latching, snapping, and clasping. These joints are not media-tight, can be designed to be releasable or non-releasable, and can be realized independently of the material. Typical applications are found in housing covers, plastic containers from the household sector, or automotive interior parts.

Force-fit or friction-fit connections are characterized by the normal stress applied to the contact surface of the joining partners, through which shear stresses can be transmitted by friction. Screw or press-fit connections are examples of this type of joint, whereby screwing always also involves a form fit. Since plastics exhibit viscoelastic properties, there is a risk of creep, which can cause stress relaxation and thus failure of the connection. The connections are in principle releasable.

In the case of material-bonded connections, the bond is formed by molecular interactions between the bonding partners. This requires close contact between the bonding partners, which is achieved by liquefying one partner. In adhesive bonding, the adhesive represents the liquid component; in welding, coextrusion, or multi-component injection molding, at least one of the bonding partners is present as a melt. A prerequisite for the formation of the bond is the compatibility of the bonding partners. The connections cannot be separated non-destructively and are frequently realized over a large area.

Due to their strong cohesion, material-bonded connections are popular in plastics technology. The molecular bond formed in this way can not only ensure the transmission of forces but can also particularly effectively ensure heat and mass transport across the connected components. The bonded surfaces range from localized weld or adhesive spots to large-area bonded film layers.

4.6.2 Challenges in Recycling

With regard to the separation of plastics by type, composites made of several plastics or of plastics combined with metal, ceramics, textiles, or paper present a challenge. Depending on the type and design of the composites, however, the components can in some cases be separated from one another and material recycling can be ensured. In this context, it is useful to classify the three types of composite described in Section 4.6.1 into releasable and non-releasable joints. Releasable connections include, for example, screw connections and some snap-fit connections. Non-releasable connections are generally material-bonded connections and in some cases also snap-fit and riveted connections.

Pre-treatment and disassembly is an intermediate step in waste treatment after waste collection and before the actual processing. Disassembly serves not only to remove hazardous substances but also to extract components that are to be recycled or recovered separately. In some cases this is required by law, such as in the scrapping of vehicles, where large plastic components such as the dashboard or the center console are dismantled. Otherwise, disassembly operations followed by targeted processing are limited to particularly high-value components, which are frequently of metallic origin [MG16, p. 29].

In addition to non-destructive disassembly, for example by releasing screw connections, partially destructive and destructive disassembly processes are also used. When individual connection points are deliberately destroyed, this can be referred to as partially destructive disassembly, in which the component itself remains intact. Destructive disassembly serves to separate non-releasable connections with destruction of the entire component [Ehr04, p. 35].

The targeted disassembly of components is characterized by individually tailored dismantling techniques with the specific plant technology required for this purpose. In this way, for example, the separation of adhesive or riveted connections can be realized in a disassembly line, which has been state of the art for more than thirty years, e.g. [Käu87, Rot09, p. 59 ff.]. A prerequisite for the economical operation of such a line is the availability of correspondingly high quantities of waste from the same component or component type, so that the investments can be amortized [MG16, p. 30 f.]. However, many multi-component products do not occur at the required frequency at waste disposal facilities, so that targeted disassembly is not worthwhile. Instead, these waste fractions are comminuted in the course of the processing steps and prepared for further treatment, which in many cases is energy recovery.

In order for material recycling to succeed instead, it is necessary to separate the composites, which, depending on the type of composite, is based on different mechanisms. In the case of form-fit plastic-plastic connections, the bond can be released by the comminution process. As soon as the flank of the component forming the undercut yields, the connection is separated. This is followed by sorting using the usual separation techniques (see 3.2.2). The same applies to friction-fit connections with plastic partners, in which comminution severs the prestressed state and thus causes the connection to lose its hold. If one of the bonding partners consists of metal, either separate disassembly or the comminution process is carried out. In the subsequent sorting process, the metal is removed from the waste stream by appropriate separators.

Since the material bond retains its bond strength even without an undercut or prestressed state, comminution can only ensure separation success to a limited extent. When comminuting large-volume parts with few material-bonded, small-area connection points, only a small proportion of the fragments contains both components. These can be sorted out and recovered separately. In the case of large-area connections such as multilayer films, however, each fragment contains all the plastics used in the component, so that comminution cannot achieve composite separation. This means that for these waste fractions, either only energy recovery remains, or the joint processing of the bonding partners as a kind of polymer blend into new products. In many cases, however, inferior material properties are to be expected, particularly when a large number of different plastics are involved whose chemical compatibility is only limited or not present at all.

Through research activities and company developments, processes have emerged in recent years that serve to release the material bond. These activities and own investigations into composite separation will be discussed in the further course of the chapter. It should be noted that by establishing such processes, the recyclability of multi-component components can be ensured. It is therefore of great importance that the requirements for the certification of recyclability must always be adapted to the current state of the art in the field of sorting and processing. However, the economic viability of the processes should be kept in mind; otherwise, only a theoretical or technical recyclability would be given (see the discussion in Section 4.3).

4.6.3 Fundamentals of Material-Bonded Joints

The material bond is based on atomic or molecular interactions between two phases. The region in which these interactions come into play is called the interface or interphase. The totality of the interactions that occur is described by the term adhesion, although despite numerous attempts there is no universally valid explanation for this phenomenon. According to [VP91], adhesion describes the phenomenon of cohesion between two surfaces that have been brought into contact with each other. This general, non-cause-related definition is due to the fact that the fundamental causes of adhesion are derived in the literature through various theories. Each of these theories considers a part of the phenomena found in experimental investigations, but none explains them comprehensively. An overview of the models that each partially describe adhesion is presented in Figure 4.9. First, a distinction is made between mechanical and specific adhesion. The latter is further subdivided into chemical and physical interactions. In the following, the basic assumptions and derivations for describing adhesion will be presented for each of the theories.

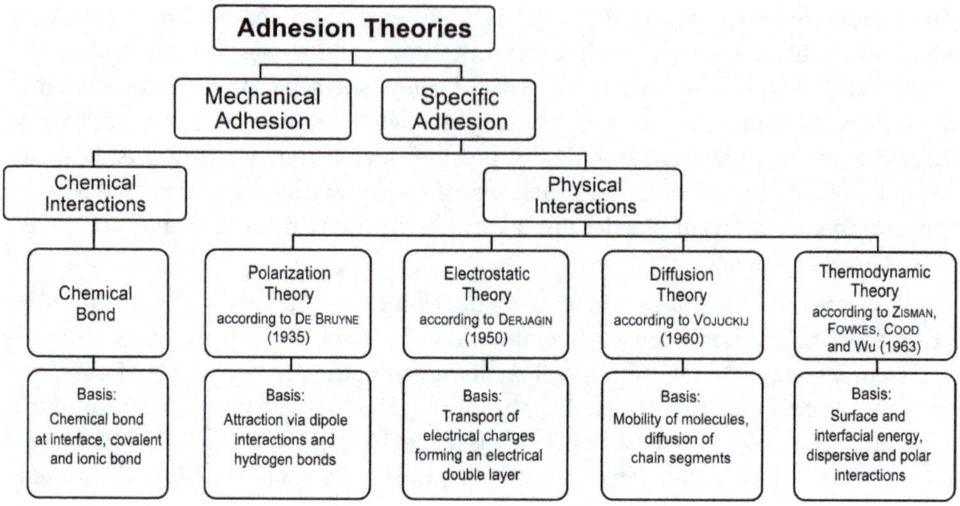

Figure 4.9 Overview of adhesion theories, after [BP83, p. 27]

Mechanical Adhesion

This type of adhesion refers to the mechanical interlocking of the bonding partners in the interphase. This occurs through structures on the surfaces such as pores or protrusions, which in the best case create an undercut in the bond and thus a kind of microscopic form fit.

Mechanical adhesion is used, for example, in the metallization of ABS. The surface of an ABS component prepared for electroplating exhibits a highly porous, cavern-like surface. This is achieved by an etching process in which the butadiene component is removed from the surface of the ABS. The metal layer deposits in the pores and provides anchoring of the metal in the ABS component, which is also called "press-stud anchoring". In the same way, adhesives can also penetrate into the pores and produce the same effect. Effective separation of the components is possible by liquefying one of the components [Pot04, p. 74].

In some cases, strong roughening of the surfaces is also used as a pre-treatment for adhesive bonds to supposedly enhance mechanical adhesion. However, this is controversial, because roughening can simultaneously also cause an enlargement and functionalization of the surface [Bis88, p. 10].

Specific Adhesion

The term specific adhesion encompasses the theories that focus on the physical and chemical conditions in the interphase as explanatory approaches. Interactions at the atomic or molecular level can be described as bonding forces at the interface between two bonding partners.

The bonding forces that occur can be divided into chemical and physical bonds. Chemical bonds are primary valence bonds and can be ionic, metallic, or covalent bonds. In the case of polymers, covalent bonds are found in the molecular chains. Compared to secondary valence bonds, primary valence bonds are characterized by a short range and high bond energy (see Figure 4.10). The intermolecular secondary valence bonds can be subdivided into hydrogen bonds and van der Waals bonds.

The latter in turn comprise dipole-dipole bonds (Keesom interaction), induced dipole bonds (Debye interaction), and dispersion forces (London interaction). Direction-dependent dipole-dipole bonding forces occur between molecules with a permanent dipole moment. Polar molecules can induce a dipole moment in a polarizable molecule, resulting in lower bonding forces compared to the Keesom interaction. Dispersion forces arise from a statistically changing electron density in the molecules. This can temporarily lead to dipole moments and thus to the formation of bonding forces. Compared to van der Waals bonds, hydrogen bonds represent an intermolecular type of bond with the highest achievable bonding forces. They are generated by the polar bond between a hydrogen atom and a more electronegative atom. The resulting positive partial charge of the hydrogen atom forms bonds with free electrons of neighboring atoms [Mon16, p. 11 ff.].

Bond Type			Range of Bonding Force [nm]	Bonding Energy [kJ/mol]
Chemical Bonds	Ionic Bond		0.15 – 0.24	335 – 1050
	Covalent Bonds		0.15 – 0.24	63 – 920
	Metallic Bonds		0.26 – 0.3	110 – 350
Intermolecular Bonds	Hydrogen Bond		0.3 – 0.5	< 50
	Van der Waals Bonds	Dipole-Dipole Bonds Keesom Energy	0.3 – 0.5	4 – 21
		Dipole-Induced Dipole Debye Energy	0.3 – 0.5	2
		Dispersion Forces London Energy	0.3 – 0.5	4 – 42

☐ Rarely occur as intermolecular bonds in composites; present as primary valence bonds in polymer chains

☐ Bond types occurring in plastic composites

Figure 4.10 Intermolecular interactions [BP83, p. 24; Dut00, p. 31; Hab09; p. 316; Nik05, p. 12]

Chemical Adhesion Theory

When bonding metals to plastics, covalent bonds can be detected. Reactive adhesives and special pre-treatment processes are used in this context [Mon16, p. 23]. Chemically reactive adhesives such as epoxy resin or PUR adhesives react with functional groups on the surface of metals, resulting in a strong bond. Such bonds cannot be separated by solvent-based processes, since the cured adhesives cannot be dissolved. For the purpose of composite separation, such bonds are poorly suited.

Polarization Theory

De Bruyne postulated that adhesion between two bonding partners is only possible when opposite polarity is present. For the bond, polar functional molecular segments are therefore necessary, through which dipole forces are formed and hydrogen bonds are created. However, the adhesion of non-polar bonding partners cannot be explained by this theory [Mon16, p. 17].

Electrostatic Theory

Beginning with the work of Derjaguin, the theory emerged that when two materials with an initially existing potential difference come into contact, charge carrier diffusion takes place. As a result, an electrical double layer forms. The migration of charge carriers occurs from the donor to the acceptor according to the thermodynamic potential difference. Due to the attracting, opposing charges, an attractive force between the two components subsequently exists. This theory is considered in particular for combinations of metal and polymer [Dut00, p. 34 ff.; Mon16, p. 18].

Diffusion Theory

A theory limited to the bonding of two polymers assumes the formation of an interphase in the contact zone, in which molecular chain segments of the two bonding partners diffuse into each other. This requires a sufficiently high mobility of the molecular chains, which is present in the melt state. The entanglements of the molecules in the interphase can be regarded as mechanical interlocking at the molecular level. The theory can explain the adhesion between non-polar plastics. Since diffusion processes are always temperature-dependent, the interdiffusion of molecules can proceed more rapidly when the temperature is increased. The diffusion theory is used in particular in the welding of plastics, in conjunction with relaxation processes of neighboring molecules after joining. In composite injection molding, on the other hand, diffusion of the molecules is only possible to a limited extent, since the first injected component has already solidified when the second component is overmolded. However, through heat conduction at high melt temperatures, the near-surface regions of the first component can be heated to such an extent that diffusion processes can still take place [Mon16, p. 17; Pot04, p. 73].

Thermodynamic Theory

The basis of this theory is wetting as a prerequisite for adhesion between two phases. The wetting behavior depends on the so-called surface energy of the materials, which can be illustrated as follows: within the interior of a homogeneous substance, all bonding forces are saturated by neighboring molecules, whereas this does not apply to the molecules located at the surface. They therefore possess a free surface tension or surface energy [BP83, p. 16, Pot04, p. 44]. The surface molecules can subsequently build up adhesion with other substances by exerting attractive forces at the molecular level. If two bonding partners are adhesively bonded to each other, the so-called work of separation, or work of adhesion W_A must be expended for composite separation, which can be calculated using the Dupré equation.

$$W_A = \gamma_s + \gamma_l - \gamma_{sl}$$

where γ_s is the surface tension of the solid, γ_l is the surface tension of the liquid, and γ_{sl} are the interfacial energies. How well adjacent materials form bonds can be assessed via the specific wetting behavior [Pot04].

Experimentally, wetting can be quantitatively captured using the so-called contact angle measurement, and by applying Young's equation a relationship between the interfacial energy and the surface tensions of the materials involved can be established (see Figure 4.11). Young's equation with the contact angle θ reads:

$$\cos\theta = \frac{\gamma_s - \gamma_{sl}}{\gamma_l}$$

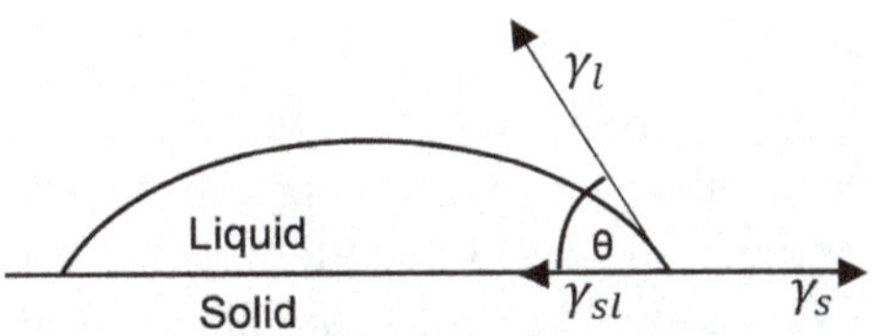

Figure 4.11 Liquid droplet on a solid

By rearranging and substituting Young's equation into the Dupré equation, the Young-Dupré equation is obtained:

$$W_A = \gamma_l(1 + \cos\theta)$$

It can now be derived that at a contact angle of $\theta = 0°$, i.e., with complete spreading of the liquid droplet, the work of adhesion becomes maximum. In this case, the interfacial tension is minimized and the work of adhesion equals the work of cohesion. The latter must be expended to separate a single-phase system [Gut91, p. 118].

The interfacial tension is a measure of the physicochemical interactions that the two composite partners in contact exert on each other. In order to account for intermolecular interactions in a cause-specific manner, various interaction parameters have been introduced and the surface tension has been split into a polar and a dispersive component. The quantification of these components allows for interpretations regarding the polarity and the potential adhesion tendency of the composite partners. However, numerous studies in the past have shown that no general relationship between surface energy and adhesion strength can be established. Good wetting is in many cases a necessary condition for good adhesion, but not a sufficient one. The relationships described apply only to homogeneous materials and ideally smooth surfaces. Therefore, this theory is also not suitable for describing adhesion in its entirety [Mon16, p. 22; Nor18; Ras12].

Summary

As outlined, major challenges arise in predicting adhesion from a molecular perspective. To date, there is no universally applicable method for calculating adhesion strength from the theories presented. Adding to this difficulty is the fact that adhesion strength cannot be quantified experimentally either. The theoretical adhesion strength σ_H resulting from the sum of interactions is defined according to [BP83]:

$$\sigma_H = \frac{F_i}{A_w}$$

It is the quotient of the internal force F_i, which cannot be determined by mechanical testing, and the true surface area $A_w$that is effective for adhesion and arises during purely adhesive fracture. In experiments, only the bond strength can be determined as a sum of the effects responsible for the cohesion of the composite. It is determined via the externally measurable force F_a and the geometric surface area A_g.

$$\sigma_V = \frac{F_a}{A_g}$$

The experimental determination of bond strength is based on various testing methods and depends on numerous factors. These concern, e.g., the geometry of the test specimens, the testing methodology, the test parameters, the production of the test specimens, or the ambient conditions. The values obtained are either qualitative and comparative in nature, or quantitative and mechanically justified [BP83, p. 243].

4.6.4 Mechanisms of Bond Separation

After Section 4.6.3 presented the fundamental theories on the formation of a material bond, the following section discusses approaches to how such a bond can in principle be separated again. The relevant intermolecular interactions and possibilities for directly loading the interface will be addressed. Unless otherwise noted, the composites under consideration are polymer/polymer composites.

Figure 4.12 shows two mechanisms of bond separation that aim to reduce secondary valence forces and thus act at the molecular level. Solvents act as new bonding partners for the free valences of the molecular chains. An increase in temperature increases the distance between the chains, thereby reducing the bonding forces.

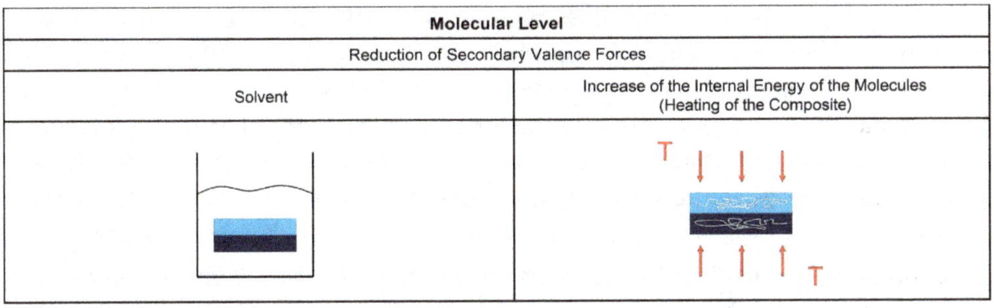

Figure 4.12 Mechanisms of bond separation at the molecular level

4.6.4.1 Use of a Solvent

By using a solvent, the secondary valence forces between the molecular chains of one of the composite partners are reduced to such an extent that the polymer chains are present individually in the solvent. Thus, the secondary valence forces between the composite partners are not primarily reduced; rather, separation occurs through complete dissolution of one of the composite partners. The undissolved component can be removed and the dissolved polymer can be recovered by removing the solvent. In xylene, for example, PP, PS, LDPE, HDPE, and PVC dissolve at different temperatures, whereas PET is insoluble in xylene.

When selecting the solvent, attention must be paid to the so-called solubility parameter, which should be similar to that of the plastic. This solubility principle is also described by the phrase "like dissolves like." The Hildebrand solubility parameter δ is defined as the square root of the quotient of the cohesive energy E and the molar volume V [MK03, p. 11]:

$$\delta = \sqrt{\frac{E}{V}}$$

The quotient E/V is also called cohesion energy density and describes the energy required to break all intermolecular, physical interactions in the volume. Since the intermolecular binding forces are based on different effects, Hansen defined three solubility parameters. The first describes the nonpolar interactions (dispersion forces) due to the free movement of electrons around the atoms and is called the dispersive solubility parameter. The second is the polar solubility parameter, which is based on dipole-dipole interactions. The third in this series takes hydrogen bonds into account. According to Hansen, the following applies when incorporating the dispersive (d), polar (p) and hydrogen (h) bond energies:

$$\frac{E}{V} = \frac{E_d}{V} + \frac{E_p}{V} + \frac{E_h}{V}$$

respectively

$$\delta^2 = \delta_d^2 + \delta_p^2 + \delta_h^2$$

The Hansen parameters are fully documented for solvents, while they must be determined with considerable effort for plastics. The methods for directly measuring the individual parameters are not applicable to plastics for various reasons, e.g. because the substance to be examined must be converted into the vapor phase and plastics are already decomposed at that point. An alternative is the determination of the parameters through a large number of solubility tests. The parameters of the solvents are known, and if the plastic dissolves well, the parameters can be inferred [Woi14, p. 25].

The dissolution process is determined on the one hand by the penetration of the solvent and on the other hand by the extraction of the molecular chains from the polymer (see Figure 4.13). According to Überreiter, five layers form on the surface of the polymer in solution [Ueb68]. The solvent has penetrated as far as the infiltration layer and fills the finest voids in this layer that have formed due to defects in the polymer structure. In the swollen layer, the solvent has already caused swelling of the polymer. However, it is still in a glassy state. In the swollen gel layer, the secondary valence forces have decreased significantly and the material behaves in a rubber-like manner. The solvent forms bonds with the free valences of the molecular chains and thus takes the place of the neighboring molecular chain that was previously available as a bonding partner. In the liquid layer, the polymer chains are completely surrounded by the solvent [MK03].

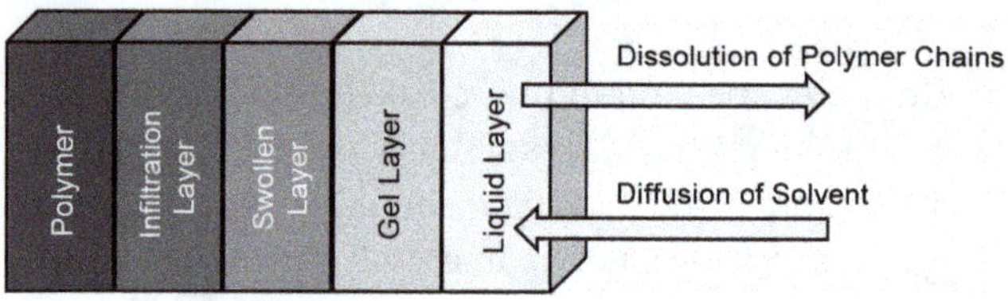

Figure 4.13 Schematic structure of the surface of a polymer in contact with a solvent according to [ZLN18]

The extraction or disentanglement of the polymer chains from the molecular network takes longer with higher molecular weight. A regular chain structure and high crystallinity delay the effect of the solvent due to the high secondary valence forces. An elevated temperature during dissolution, on the other hand, improves solubility [MK03].

Selective dissolution is only possible for thermoplastics, because in thermosets and elastomers the crosslinking sites prevent dissolution. The bonding partner can consist either of another plastic or of a different material. If it is also a plastic, the solvent must be selected such that this material is not dissolved as well.

Although selective dissolution as a process for composite separation has been known since the 1990s, there has been no industrial implementation on a large scale to date. The current developments on the part of industry are presented in Section 4.6.4.1.

4.6.4.2 Increase of the Internal Energy of the Molecules

The second approach to reducing the secondary valence forces is achieved by supplying energy to the composite. If the internal energy of the molecules increases, the Brownian molecular motion is intensified and consequently stronger molecular vibrations occur. These lead to increased collisions between the atoms of neighboring molecules and thereby to a greater distance between the intermolecular bonding partners. This has a direct influence on the magnitude of the binding forces between the molecular chains [RHB12, p. 261]. Calculations of the interaction energies of various atomic constellations have shown that the magnitude of this energy is inversely proportional to the sixth power of the distance. Keesom, for example, obtains for two free dipoles with dipole moments $\vec{p_A}$ and $\vec{p_A}$, R as the distance between the dipoles the following expression for the mean energy [BP83, p. 52]:

$$\overline{U} = -\frac{2\,\vec{p}_A^{\,2}\,\vec{p}_B^{\,2}}{3(4\pi\varepsilon_0)^2 kTR^6}$$

With: k = Boltzmann constant$T$ = absolute temperature

A similar expression applies to the energy between permanent and induced dipoles, but without the temperature dependence [BP83, p. 53]:

$$\overline{U} = -\frac{\vec{p}_B\vec{\alpha}_0(A)\,\vec{p}_B}{(4\pi\varepsilon_0)^2 R^6}$$

With: = polarizability of molecule A

Bischoff and Possart give, as a first approximation for two molecules, an energy as the sum of the i atoms of molecule A with j atoms of molecule B [BP83, p. 55]:

$$U_{AB} = \sum_{i\epsilon A}\sum_{j\epsilon B}\frac{C_{ij}}{R^6}$$

For the constant C_{ij} various expressions are cited in the literature according to the conditions in the molecule, which can be substituted into the formula. However, the dependence of the energy on the reciprocal distance to the sixth power is always present in these calculations. To return to the argument from the beginning, it becomes clear that an increase in the distance between the molecules brought about by the addition of heat leads to a strong reduction in the interaction forces between the molecular chains [Kai16, p. 72; DHM22, p. 74].

The reduction of the secondary valence forces results in bonds being more easily broken and reformed, and, upon reaching the glass transition or melting temperature, sliding of molecular chains becomes possible. In conjunction with the application of external forces, it is to be expected that a composite of two polymers can be separated more easily due to the aforementioned relationships. The temperature increase and the greater distance between the molecules result, in addition to the effect of reduced secondary valence forces, in thermal expansion in the plastic, which induces shear stresses in the interface of the composite (see Section 4.6.4.4).

The intermolecular interactions with another material acting in the interface, which lead to adhesion, are generally lower than those that produce cohesive strength. It is therefore to be assumed that the adhesion forces at elevated temperature are lower than the cohesion forces. If this is not the case, however, one approach to separation may consist of converting one of the components into the melt state and filtering the solid component out of the material stream. The secondary valence forces are thus so low that the cohesive strength becomes minimal. The composite separation by application of mechanical input at a specific temperature favorable to composite separation is described in a US patent by Tucker et al. from 2006 [TM06].

The energy input into the composite can be achieved by convective heat transfer to the surface and subsequent heat conduction through the composite partners to the interface. Another possibility involves the use of microwaves and a polymeric interlayer between the composite partners, which contains induction-affine fillers.

While the mechanisms for separating a composite shown so far directly target the secondary valence forces that significantly influence adhesion, the following section deals with overcoming the composite strength through external or externally induced internal forces (see Figure 4.14). In this context, the application of an external force can also be combined with the possibility just presented of reducing the secondary valence forces by heating the composite.

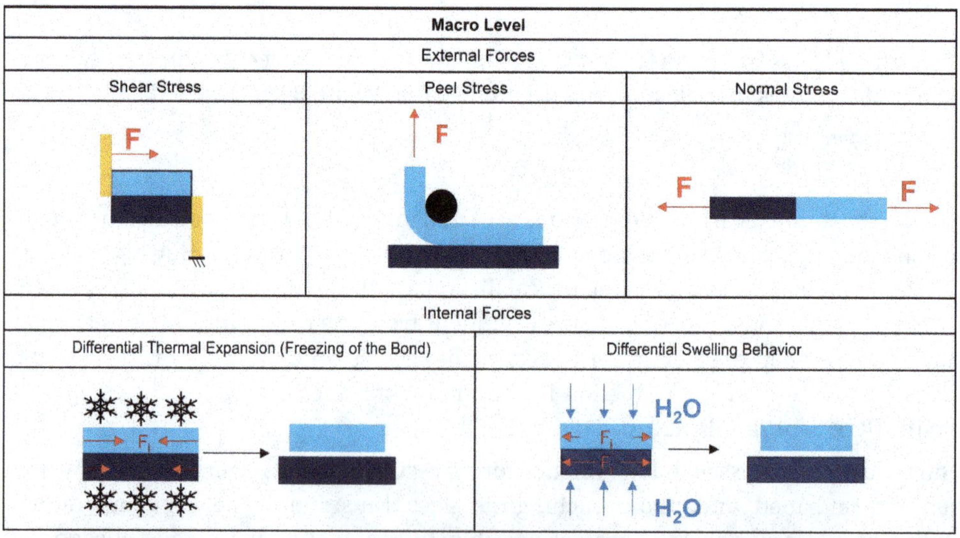

Figure 4.14 Mechanisms of composite separation at the macro level

4.6.4.3 Shear, Peel, or Normal Stress

Overcoming the composite strength through external forces requires a sufficiently high cohesive strength of both composite partners. Depending on the geometry of the bond area, shear, peel, or normal stress is suitable for separating the composite.

It must be noted that the application of an external force can induce higher stress in the material itself than at the interface. For example, this is the case when, under normal stress, the interface is larger than the cross-section of one of the composite partners. In that case, failure of the component can occur before adhesive failure takes place. The presence of a multiaxial stress state in the component due to the applied force can also lead to "premature" fracture of the component.

In principle, peel stress is to be preferred when selecting the force application method, because in this case the force is introduced along a line, meaning that only a small portion of the bond area transfers the force. High stress peaks are generated in the interface, which can be conducive to adhesive failure [Hab09, p. 533]. However, it must be noted that a peeling load, as shown in Figure 4.14, can only be applied to composites in which one of the components has sufficient flexibility.

Pure shear stress requires high forces for large bond areas, which is why it must be considered unsuitable for composite separation. However, for a mechanical implementation of composite separation of comminuted multi-component fragments, shear stress is the easiest to realize. If only small bond areas are present and the composite has already been weakened by suitable measures, composite separation by shearing can also be successful. Furthermore, depending on the separation principle of the machine, components of peel and normal stress always occur in addition to shearing.

Normal stress is characterized by the force acting perpendicular to the bond area. It is therefore, like peel stress, a lifting force, but one that is introduced over an area rather than acting along a line.

4.6.4.4 Internal Forces Due to Different Thermal Expansion

In addition to externally applied forces, internal forces can be generated by two approaches, both of which aim to build up a shear stress at the interface that is sufficiently high for composite separation. The first approach consists of exploiting the different thermal expansions of the two composite partners. Plastics have a significantly higher thermal expansion compared to metals. This can be expressed by the linear thermal expansion coefficient α, which is defined with the initial length L_0, the partial length $\square L$ and the partial temperature $\square T$ as follows [DHM22, p. 365]:

$$\alpha = \frac{1}{L_0} \times \left(\frac{\partial L}{\partial T} \right)$$

As temperature increases, the distance between molecules increases, causing the specific volume and thus the length of the component to increase. In addition to this thermal, reversible volume change, the following further effects occur according to [DHM22]:

- Reversible transformations (e.g., melting processes)
- Reversible transitions (e.g., glass transition)
- Irreversible volume changes due to aging processes (e.g., relaxation, post-condensation)

As can be seen from Figure 4.15, the thermal expansion coefficient increases with temperature, in some cases doubling over a temperature interval of 30 °C. At room temperature, polyethylene shows a high value of approximately $18 \times 10^{-5}\ K^{-1}$ compared to other plastics. Corresponding values for metals are below $2.5 \times 10^{-5}\ K^{-1}$. The sometimes abrupt increase in values in plastics is attributable to the transitions or transformations already mentioned.

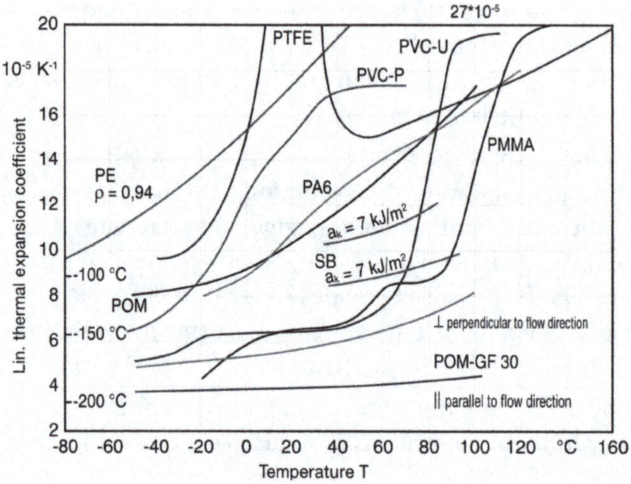

Figure 4.15 Dependence of the thermal expansion coefficient on temperature [BBO+13, p. 128]

If the change in length is suppressed, residual stresses develop in the plastic. This is the case when two plastics in a multi-component part are bonded together by material bonding and their thermal expansion coefficients α differ significantly. The suppressed change in length of the plastic with a high α leads to a shear stress at the interface that can overcome the composite adhesion. In the course of the separation process, the composite is therefore heated or cooled, thereby generating the shear stresses that lead to separation. With regard to plastics, cooling down to sub-zero temperatures is useful in this context. The advantage over high temperatures is that brittle material behavior is simultaneously present and the components can be separated by liberation comminution (see section 4.6.5). Furthermore, at high temperatures, molecular rearrangements occur again, leading to relaxation of the intentionally introduced stresses. It should also be noted that residual stresses are already introduced during the manufacturing process due to suppressed shrinkage. These would tend to increase further upon additional cooling, whereas an increase in temperature would initially eliminate the residual stresses. Liberation comminution with upstream cryogenic treatment is state of the art in metal/plastic separation, e.g., for printed circuit boards, transformer laminations, or other electronic components with plastics as insulating elements. An advantage here is that the thermal expansion coefficients of metals are in some cases up to ten times lower than those of plastics [DHM22, p. 366; MG16, p. 177]. In recent studies on the separation of structural adhesive bonds, it has been shown that the embrittlement of an epoxy adhesive at –55 °C leads to a significant reduction in impact peel strength [CMB+22].

4.6.4.5 Internal Forces Due to Different Swelling Behavior

In addition to the volumetric expansion of plastics due to heat, the penetration of liquids into the polymer structure also causes a change in volume. This process is called swelling, which is fundamentally a precursor stage during the dissolution process of plastics. Here, however, the plastic is not to be dissolved, but rather the build-up of a shear stress is to be induced by swelling forces. For this purpose, solvents can be used that show only a low compatibility with the plastic in question with regard to solubility, which can be determined in individual cases through elaborate laboratory tests. If the solubility parameters match very well, the undesired dissolution of the plastic would occur. A more ecologically sensible and technically less complex method is the use of water as a swelling agent. Polar plastics show an increased affinity for binding water molecules in their molecular structure through hydrogen bonds. This leads, in addition to changes in mechanical properties, to an increase in volume. Non-polar plastics such as PE or PP absorb little to no moisture, while polar plastics such as PC or PET and in particular PA show a high water absorption capacity. In polyamides, the increase in length can be 0.3% per 1% water absorption, whereby polyamide 6, for example, can absorb over 10% water. The high water absorption of this type of plastic is attributable to the amide groups in the polymeric structure [Kai13, p. 377]. If a polyamide is combined in a composite with metal or with a plastic that does not have a high water absorption capacity, the different swelling behavior leads to a shear stress at the interface due to the constrained increase in length. It should be noted, however, that the water in the polymer structure also leads to a reduction in strength and stiffness [DHM22, p. 538; BBO+13, p. 162 ff.].

4.6.5 Process Engineering Methods for Composite Separation

As already explained, component-specific disassembly lines are used only in a few cases to separate composites (see Section 4.6.2). Prerequisites for commercial plant operation are the availability of larger quantities of waste and the possibility of recovering high-quality materials. The processes presented below therefore concern component-independent processing steps aimed at separating the bonds. Economically viable technologies for composite separation of such multi-component, heterogeneously present waste for material recycling can in principle be applied in the following three phases of processing:

- During the comminution process
- During the processing of the comminuted waste fragments
- In the remelting phase of the plastics

In which phase the separation can be achieved most efficiently depends in particular on the following factors, derived from the discussion of the mechanisms of composite separation:

- Type of bond (form-fit, force-fit, material bond)
- In the case of material bonding: level of bond strength
- Geometry and size of the interface (point-like, planar)
- Materials of the bonding partners (different plastic types, metal, ceramic, textile or paper)
- Thermal properties of the bonding partners (glass transition, softening, melting and decomposition temperature, thermal expansion behavior)
- Mechanical properties of the bonding partners (strength, hardness, stiffness)
- Solubility and water absorption capacity of the plastic types

Before the respective material stream of multi-component components can pass through the separation process, it must first be collected and sorted by composition. Collection and sorting must be carried out according to the requirements of the respective separation processes. Ultimately, two principles can be distinguished in sorting for the separation processes:

- Component-specific collection/sorting

  •For some products, the same or at least similar materials are used across manufacturers, so that the separation process can be tailored to the material combination of these components. Separate collection for these products leads to efficient logistics without additional sorting effort. Examples of such products are CDs or DVDs as well as printed circuit boards.

- Material-specific collection/sorting

  •For some processes, sorting tailored to a specific material or material combination may be useful. If, for example, a process is aimed particularly at metal-plastic separation, such products can be specifically discharged in corresponding sorting facilities. A further example concerns multilayer films containing PA or PET fractions. In current studies, it has been demonstrated at least in a laboratory environment that the composition of multilayer films can be analyzed by NIR spectroscopy

[CKF+20; CKW+21]. In this way, the mixed plastics fraction can prospectively also be post-sorted from the sorting facilities for lightweight packaging.

Since pre-sorted material streams are an absolute necessity for some of the processes listed below, sorting technology is indispensable in the development of new processes. However, sensor-based methods are already at a stage in the research environment where they can be used on an industrial scale as soon as economically viable downstream processes for further processing exist. In the following, selected options for composite separation will be presented, some of which have already been state of the art for some time, while others are still the subject of current research.

4.6.5.1 Liberation Comminution

In the course of comminution, in addition to reducing the particle size of the waste, the liberation of composites can be achieved. The comminution energy introduced into the waste pieces stresses the respective bond point or the interface of the composites and leads to separation of the composite in the respective waste fragments. Impact stresses are used as the type of loading for brittle waste, while for non-brittle waste, according to [Sch02], tearing stresses are more commonly applied. The tearing stress is a mixture of tension, bending and torsion, which can be realized by means of hammer shredders. For force-fit and form-fit connections, composite separation is achieved by comminuting the load-bearing elements of the bond points. For material bonds, on the other hand, it is necessary for the separation to occur through mechanical loading of the interface between the components. If the separation of material bonds is not achieved, comminution takes place into very small particle sizes, so that the ground material contains one fraction of each of the two composite partners as well as a mixed fraction.

So that the supplied energy is not expended primarily on elasto-plastic deformation, cold embrittlement through the use of liquid nitrogen may be useful for ductile or rubber-elastic materials such as thermoplastics or elastomers. The liberation of composite structures made of high-strength steels with fiber-reinforced plastics, which are joined mainly by point-like bond points, is readily possible using a hammer mill. However, the wear of the hammers is not negligible [LKS17].

4.6.5.2 Solvent-/Chemical-Based Processes

As already shown in Section 4.6.4 in the presentation of the mechanisms for composite separation, the use of a solvent reduces the secondary valence forces so that the polymer in question goes into solution. Initial work on this dates back to the USA at the beginning of the 1990s and resulted in a US patent that has since expired [NL94]. In a composite of several plastics, one component after another goes into solution at different temperatures and the target polymer is filtered out. An overview of further patents and developments over the last thirty years can be found in [KSS17].

Two processes already used industrially in Germany are the CreaSolv®and the Newcycling process. The latter is used by the company APK AG from Merseburg to produce polyethylene and polyamide granulates in pure grades from mixed plastic waste containing multilayer packaging. After comminution, cleaning and density separation, contact with a solvent takes place. This is followed by separation processes between solid and liquid components, polymer purification stages and removal of the solvent. Using this technology, the company processes an input capacity of 8,000 t/a [NN02-ol]. At the end of 2021, according to company information, the decolorization of LDPE from household collection waste was also achieved [NN03-ol].

Designed for significantly more plastics is the CreaSolv®process of the Fraunhofer Institute for Process Engineering and Packaging IVV in Freising. Creacycle GmbH from Grevenbroich supplies the solvents, e.g. for the extraction of polyolefins, polystyrene or PET from packaging. Special applications include the recovery of styrene copolymers from waste electrical and electronic equipment, polyamides and PBT from post-industrial composite waste, as well as expanded polystyrene (EPS) from insulation materials. The type of solvent used in each case is not disclosed. The process has been under further development for approximately twenty years and, following several successfully completed pilot projects, is on the verge of establishing a concept for a commercial plant [SFM+20; SNK+22].

In general, solvent-based processes face challenges regarding the large quantities of solvents required, their flammability, the high vapor pressure required, the high costs for the equipment technology, and in particular the toxicity of the solvents [MG16, p. 289].

In addition to selective dissolution, i.e. the complete dissolution of a plastic, there is also the possibility of achieving delamination of layers by means of chemicals. The process developed by the company Saperatec GmbH from Dessau operates according to this principle. Comminuted waste is brought into contact with water-based chemicals tailored to the composite, and delamination of the layers takes place under heat and a stirring process. A screening process returns the separation liquid and the separated products are subjected to a washing process. The individual components can then be sorted. According to company information, composites of plastic and aluminum as well as plastic and glass can be separated. A commercial plant with a recyclate output of 16,000 t/a is currently being built in Dessau, which can separate PET/aluminum/PE, PET/aluminum, aluminum/PE and PE/aluminum/PE composites and is scheduled to go into operation in 2023 [NN04-ol; NN05-ol].

4.6.5.3 Weakening of the Composite by Temperature Increase

In the following, results of investigations into composite separation are presented, which are based on the increase of internal energy through heat supply. The temperature increase achieves the following effects, which influence the composite separation:

- Reduction of the secondary valence forces between the molecules

  The cohesive force within the two components decreases, so that the material strength and stiffness decline. However, the actual separation effect is intended to be generated through the reduction of the binding forces in the interface of the composite. If the cohesive forces become very low, failure of one of the components can occur before the decrease in adhesive forces leads to a separation of the composite.

- Thermal expansion of the composite partners

  Due to the different thermal expansion behavior of the composite partners, a shear stress occurs in the interface layer.

- Stress relaxation

  Residual stresses introduced into the composite during the manufacturing process or due to differential thermal expansion during the separation process can relax at elevated temperatures as a result of greater chain mobility. This causes the stresses acting in the interface layer to decrease. Depending on the temperature and the type of plastic, the softening, glass transition, post-crystallization, and melting ranges are passed through. What effects these have on the residual stress situation in connection with the processing history is subject to case-by-case consideration.

In addition to increasing the temperature of the composite, applying a force at the interface is required to separate the components. As in Section 4.6.5.1 executed, the shear stress is well implementable by machine, however the peel stress is more effective with regard to a concentrated force introduction.

By means of a temperature between 50 and 110 °C, the approach presented reduces the composite adhesion to such an extent that the composite is separated by the introduction of shear energy using a stator-rotor unit. When designing the joint and the associated material selection, the designer is in most cases concerned with achieving the best possible adhesion. In hard/soft combinations of a thermoplastic elastomer (TPE) and a thermoplastic, for example, the TPE soft components are chemically adapted for adhesion to the corresponding hard components [ZSL21]. Process engineering measures are also applied in coextrusion, composite injection moulding, or in bonding or welding, with the process parameters being optimised to ensure that the components adhere to each other as well as possible. Particularly in the field of composite injection moulding, numerous studies have been published dealing with the optimisation of adhesion, e.g. [Kuh98; Küh05; Gri10; Kle11; Hop13; MHK13; Nor18]. The goal of providing pure material streams at the end of the life cycle is not pursued, or is pursued only inadequately. Yet this aspect in the product development process can help to make exactly this possible ("Design for Recycling"). For many applications, the maximum achievable adhesion forces are not necessary at all, especially in the case of large-area applications, such as overmoulding of large surfaces in automotive interiors, or multilayer pipes. The requirement for high composite strengths at elevated temperatures is also not relevant for many applications. The material selection and knowledge of composite adhesion at elevated temperatures offer the possibility of exploring the potential for composite separation and thus for achieving recyclability.

In the following, some combinations will be presented with regard to their separability at elevated temperature, which are typically used as multilayer composites. Material parameters important for the analysis of the results can Table 4.1 be taken.

For the combinations of two thermoplastics, hereinafter referred to as hard/hard combinations, the testing of bond strength is carried out using the specimen geometries shown in Figure 4.16 under normal (a), shear tensile (b) and peel stress (c).

For the hard/soft combinations, the peel test is carried out according to the guideline VDI2019 (d), in which the flexible tension tab is deflected by 90° for the test. The test specimens are produced by injection moulding. As explained in more detail in the further course, some investigations are carried out on a three-layer composite. The intermediate layer (shown in green in Figure 4.16) has a thickness of 2 mm for the hard/hard combinations, while it is 1 mm for the peel tests of the hard/soft combinations [HS22].

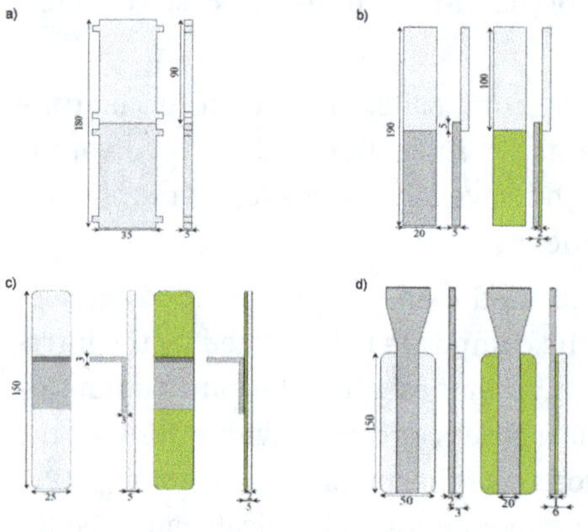

Figure 4.16 Test specimen geometries for the tensile test (a), lap shear test (b), peel test for rigid/rigid combinations (c) and for rigid/flexible combinations according to VDI2019 (d)

The evaluation of fracture patterns is carried out according to the categories shown in Figure 4.17. This allows a statement to be made as to whether it is a pure adhesive fracture without remaining residues or whether residues remain on the rigid component.

Table 4.1 Characteristic Values of the Materials Used

Material	Hardness [shore A]	Density [g/cm³]	E-modulus [MPa]	Tear/tensile strength [MPa]	Vicat softening temperature [°C]	Melting temperature [°C]
PMMA Plexiglas 8N; Röhm GmbH	k. A.	1,19	3300	77	108	117
PC/ABS Novodur H801; Ineos Styrolution Group GmbH	k. A.	1,07	2400	49	106	n/a
PC/ABS Bayblend T85XF; Covestro AG	n/a	1,14	2300	50	130	260
PA6 GF30 Ultramid B3EG6; BASF SE	n/a	1.36	9500	115	220	220
PA12 Vestamid NRG2101; Evonik Industries AG	n/a	1.02	1300	40	150	177
TPU Desmopan 9380A; Covestro AG	80	1.10	n/a	39.9	n/a	n/a
TPU Desmopan 1080A; Covestro AG	80	1.20	n/a	35	n/a	n/a
TPS Thermolast K TC6MLB; Kraiburg TPE GmbH & Co. KG	60	1.10	n/a	5.0	n/a	n/a
EBA Lucofin 1492 MHG; Lucobit AG	n/a	0.924	42	11	60	95
PP Purell HP570M; LyondellBasell N.V.	n/a	0.90	1400	33	154	165
HDPE BorSafe HE3490-LS-H; Borealis AG	n/a	0.96	1100	25	n/a	n/a
Admer NF518E; Mitsui Chemicals Europe GmbH	n/a	0.91	n/a	14	80	121
EVA Escorene Ultra UL05540EH2; ExxonMobil Chemical	44	0.96	3.3	5.3	n/a	48

In existing recycling facilities, hot washers with washing temperatures of approx. 80 °C are used, among other things, during the cleaning of plastic waste. In order to integrate the composite separation as easily as possible into existing processing steps, it makes sense to supply the energy via water as a heat transfer medium. All tests are therefore carried out at room temperature and at the temperature levels achievable for hot washers of 50, 65, 80 and 95 °C (in some cases 110 °C). For this purpose, the samples are preheated in an oven until the preset temperature is also present at the separation plane. The influence of water as a medium on the composite adhesion is not discussed at this point [HM21; HS22].

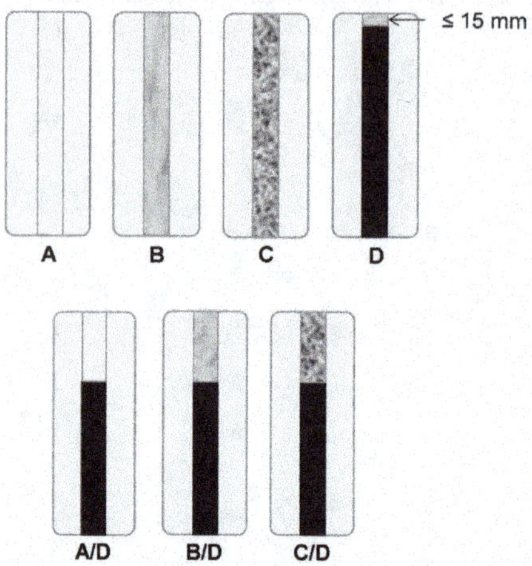

A	No residues of the TPE on the substrate
B	1 - 50 % residues of the TPE on the substrate
C	50 - 99 % residues of the TPE on the substrate
D	Fracture of the soft component (≤ 15 mm)
E	Specimen failure – residues of the substrate on the TPE (not shown)

A/D	No residues of the TPE on the substrate; fracture of the TPE component
B/D	(1…50) % residues of the TPE on the substrate; fracture of the TPE component
C/D	50 - 99 % residues of the TPE on the substrate; fracture of the TPE component

Figure 4.17 Classification of fracture patterns from the peel test according to VDI2019

PC/ABS with PMMA

The combination of the plastics PC/ABS with PMMA is used in the automotive industry, e.g. in headlights and rear lights. The material bond between the components is achieved by welding or adhesive bonding. In this case, the bond is produced by multi-component injection molding for comparability of results. The material Plexiglas® 8N from Röhm GmbH is used as PMMA, which is applied in lamp covers, displays or optical lenses. Novodur® H801 is used as PC/ABS, a material for automotive interior trim, headlight housings, seat components or door handles.

The results are shown in Figure 4.18 as maximum strength for the three investigated loading conditions — tensile, lap shear and peel loading — plotted against the test temperature. In addition, the fracture patterns are indicated on the x-axis. Based on the results from the peel tests, it can be seen that up to a temperature of 65 °C only cohesive fractures away from the separation plane occur in the PMMA component; the forces required to separate the bond are higher than the material strength. Above this temperature, both cohesive fractures and separations at the interface are achieved, but with heavy deposits of PMMA on the PC/ABS surface. It can therefore be established that an adhesive separation without deposits cannot be achieved with this combination. However, a continuous decrease in peel resistance up to 90 °C is observed, and at 110 °C a significant drop, which may be related to the fact that the softening temperature of both materials is reached at approximately 105 °C. The results of the two other loading conditions (b, c) largely corroborate these findings. In both cases, only fracture pattern C with significant material deposits is achieved at higher temperatures, and the strength does not drop significantly until 110 °C.

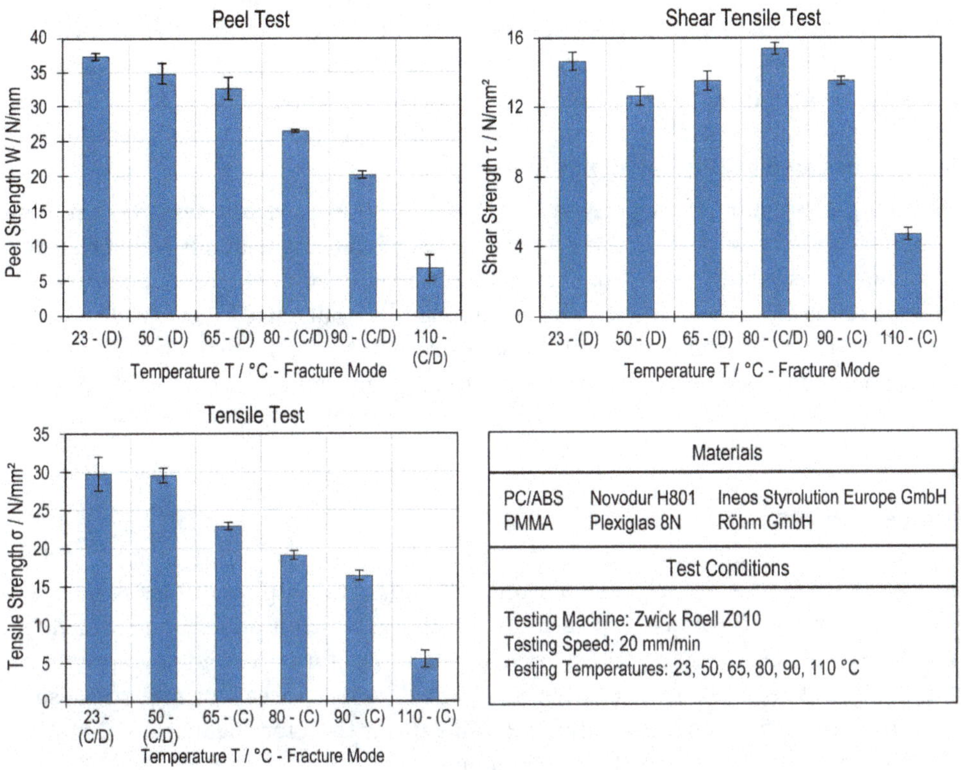

Figure 4.18 Maximum peel, lap shear and tensile strength of a PC/ABS and PMMA composite under the influence of test temperature

Overall, it can be said that this combination does not produce a purely adhesive bond separation by increasing the temperature in the composite, but retains its adhesive strength even at higher temperatures. This is quite to be expected given the location of rear lights as an example. At temperatures above 80 °C, at least a separation at the separation plane takes place, with a completely covering but thin layer of PMMA on the PC/ABS material.

PC/ABS with TPU

The rigid/flexible combination of a PC/ABS and a TPU is typically used in the automotive industry in the interior area, e.g. for so-called roller blinds in the center console [ZSL21]. The PC/ABS used, of the type Bayblend T85XF, is manufactured by Covestro, and the TPU used is Desmopan 9380A, also distributed by Covestro. This features ether-based polyol segments and exhibits lower adhesive strength to many thermoplastics than the ester-based variants.

Figure 4.19 shows the results of the peel tests for the rigid/flexible combinations as peel resistance plotted against the crosshead travel. For the combination of PC/ABS with TPU, it is clearly evident that the peel resistance decreases significantly as the temperature increases. Already at a temperature of 50 °C, the peel resistance has dropped by almost half compared to room temperature. From a temperature of 95 °C, the peel resistance is approximately 1 N/mm, which according to [ZSL21] is considered insufficient adhesion for technical applications.

With this combination, it is clearly evident that the adhesion decreases at higher temperatures, and thus a mechanical composite separation using water as a heat transfer medium, as described above, appears to be feasible. Even at room temperature, no residues of the TPU remain on the rigid component, so that a truly pure-grade separation can be achieved. The separation temperature of 95 °C is quite high for pressureless hot-wash units. In order to further reduce the composite adhesion, the use of an interlayer with the material EVA is also investigated. The results of this are presented later in the chapter.

PC/ABS with TPS

For the next hard/soft combination, the same PC/ABS of type T85XF is used, and as the TPE material, a styrene-based variant, a TPS, is now employed. This is the type TC6MLB from the Thermolast® K series by Kraiburg, which is adhesion-modified for ABS, PC and PC/ABS. It is used, for example, in handles for hand and power tools, as stop buffers, as seals, for control wheels and in all kinds of areas for realizing a soft-touch surface. In combination with the PC/ABS, this is therefore again a combination that is used particularly for technical components.

The result of the peel test shows an uneven progression over the crosshead travel. After a maximum between 50 and 100 mm, the peel resistance decreases again, followed by a continuous increase between 100 and 300 mm of travel. In general, the test is characterized by a strong decrease in the strength of the TPS material at elevated temperature. The decreasing cohesive strength leads to partial tearing of the soft component and a greater proportion of residues on the hard component at temperatures above 65 °C. Above 80 °C, good separability is observed at peel resistances <1 N/mm.

With this combination, it becomes clear that the decreasing strength of the soft component at temperatures > 65 °C means that adhesive separation of the composite is not possible. At lower temperatures, however, the bond strength is presumably too high for mechanical separation of the composite.

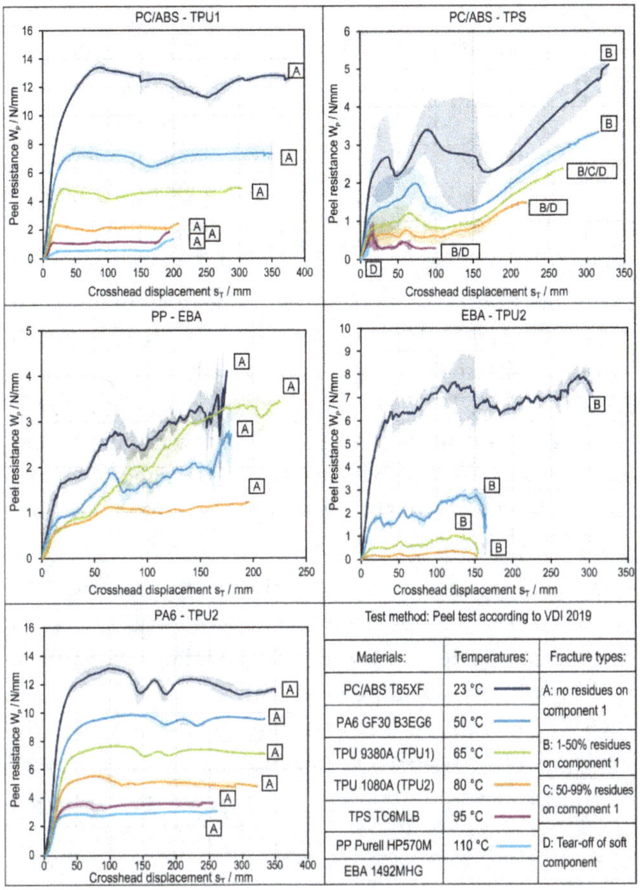

Figure 4.19 Peel curves for hard/soft combinations under the influence of test temperature (hatching represents the standard deviation)

PP with EBA – EBA with TPU

TPU as a soft component is frequently used in technical applications due to its high strength compared to other TPE materials. PC/ABS or PA6 are often used as carrier materials for large-area applications, e.g. for the production of soft-touch surfaces in automotive interiors, due to their good adhesion. However, for the mechanical requirements needed in this application, the less expensive PP would also suffice as the hard component. Adhesion to TPU is not possible or not sufficient with this plastic, however. By using an intermediate layer, the combination of PP and TPU could nevertheless be realized. Furthermore, by selecting a suitable material for the intermediate layer, not only adhesion but also composite separation can be taken into account. With the plastic EBA (ethylene butyl acrylate), which is chemically modified with maleic anhydride, composite adhesion to both PP and TPU can be expected. Due to the low softening temperature of 60 °C and the melting temperature of 95 °C, the composite separation of the entire 3-component composite of PP-EBA-TPU can be achieved by cohesive failure of the EBA component. Washing out the EBA from the PP or the TPU in a hot wash is consequently conceivable. If an adhesive failure occurs between the TPU component and the EBA intermediate layer and the EBA continues to adhere to the PP component, the EBA can be processed together with the PP to improve properties. EBA is used in many applications as an additive for impact modification or as an adhesion promoter.

To evaluate the adhesion between the intermediate layer and the two components, the 2-component variants consisting of the intermediate layer with one component each are first subjected to a peel test. The PP used as the hard component is the type Purell HP570M from the manufacturer LyondellBasell. This is in turn brought into a material bond with the EBA of type Lucofin® 1492M HG from Lucobit AG by a 2-component injection molding process. As can be seen from Figure 4.19, adhesive failure occurs at all test temperatures used, and the peel resistance decreases as expected at higher temperatures. At a temperature of 80 °C, the peel resistance is approximately 1 N/mm, so that good separability can be assumed here.

When examining the results from the peel tests of the composite of EBA with the TPU Desmopan® 1080A from Covestro AG, it becomes apparent that the peel resistance at 23 °C is significantly higher than for the composite of PP with EBA. In addition, residues of the TPU remain on the EBA. However, the mean peel resistance of this composite already drops below 1 N/mm at 65 °C, so that separation by temperature increase and mechanical input via a type of hot-wash unit is also possible here.

The results of the three-layer composite of PP, EBA and TPU are presented later in this section.

PA6 GF30 with TPU

Another common hard/soft combination consists of PA6 and TPU and is used, for example, as a caster wheel for office chairs, vacuum cleaners or similar applications. To test the separability of the composite at elevated temperatures, the PA6 Ultramid B3EG6 from the manufacturer BASF with a glass fiber reinforcement of 30% and the previously used TPU Desmopan® 1080A are used as representative examples.

The results of the peel test show that adhesive, residue-free detachment of the TPU component from the hard component is possible and that the peel resistance decreases further with increasing test temperature. However, the mean peel resistance remains above 2 N/mm even at a temperature of 110 °C, which represents sufficient adhesion for many technical application areas. Composite separation by means of a shear-intensive hot-wash unit is, on the other hand, not to be expected even at high temperatures. Initial laboratory tests with a mixing unit confirm this, which is presented at the end of this section.

Use of an Intermediate Layer

For some of the composites presented so far, composite separation is not possible due to excessively high adhesion forces. The components partially fail cohesively during the separation process, without it being possible to detach the components from one another. In the following, a further approach is presented that is intended to enable composite separation of such combinations. By using a polymeric intermediate layer with defined properties, adhesion to both components on the one hand and composite separation on the other hand are to be ensured. The use of an intermediate layer is in principle not new. In a review article, Kaiser et al. [KSS17] describe several methods for separating the composite through an intermediate layer that can be dissolved chemically, enzymatically or by means of microemulsion. There are also approaches for bonded composites, in which the adhesive is selectively dissolved either chemically by solvents or in acidic media. For PE/PET composites, separation of a bio-based intermediate layer via enzymatic hydrolysis has been investigated [CSB+16].

In contrast to this, the focus here is on an intermediate layer that ensures separation of the composite again by increasing the temperature. The intermediate layer, adapted to the material combination, is heated for composite separation, thereby reducing the adhesion to the bonding partner and allowing the two components to be separated from each other by applying mechanical energy. Another possibility for separating the composite is by softening or by complete melting of the intermediate layer. In this way too, the separation process occurs when a mechanical load is applied. For three such composites, the results of the laboratory tests are presented below. Commercially available plastics are used as the intermediate layer here. No adaptation of the intermediate layer with regard to viscosity, melting behavior and adhesion has been carried out.

HDPE with PA12 and Admer As Intermediate Layer

The investigated hard/hard combination of HDPE and PA12 with an adhesion-promoting intermediate layer of LLDPE chemically coupled with maleic anhydride is intended to simulate an application from the field of gas transport pipelines. HDPE has established itself as a well-suited material for this application due to its high ductility, low weight and comparatively low material costs. While the material exhibits high impermeability to natural gas, hydrogen molecules diffuse considerably more easily through the material. The transport of hydrogen will presumably play a not insignificant role in the context of the energy transition. Therefore, one measure to increase the impermeability of HDPE pipelines to hydrogen can consist of integrating a layer of polyamide 12 by coextrusion. Since this material is significantly more expensive than HDPE, the layer thickness should be kept small. An adhesion-promoting layer is necessary to achieve adhesion between the non-polar HDPE and the polar PA12.

The structure of the composite used in the investigations consists of HDPE BorSafe™ HE3490-LS-H from the manufacturer Borealis, the polyamide Vestamid® NRG 2101 from Evonik and the intermediate layer made of the material Admer™ NF518E from Mitsui Chemicals. The polymeric base of the intermediate layer is an LLDPE that is chemically coupled with maleic anhydride, a typical adhesion promoter. The HDPE and the PA12 are plastic types certified for use in pipelines by the trade association of the plastics pipe industry. The 3-component test specimens are produced by over-molding the pre-molded part of PA12 and Admer, previously produced in an upstream 2-component process, with HDPE. As can be seen from Figure 4.20, adhesive failure is possible at a temperature of 23 °C and the strength at 50 °C already decreases significantly under both peel and shear loading. Up to a temperature of 80 °C, failure occurs between the intermediate layer and the PA12, whereas at 95 °C the bond between the HDPE and the intermediate layer fails. The latter suggests that in a recycling process at such high temperatures, separation of all three components may be possible. The results show a strong decrease in bond strength over the investigated temperature range, particularly at the test temperature of 110 °C. Since the results of the peel test are not comparable with the results of the hard/soft combination due to the significantly different test geometry, no corresponding conclusions can be drawn. However, the extremely low shear strength at a test temperature of 110 °C and the failure occurring between the intermediate layer and the HDPE suggest that separation of the composite by a unit operated with hot water is possible.

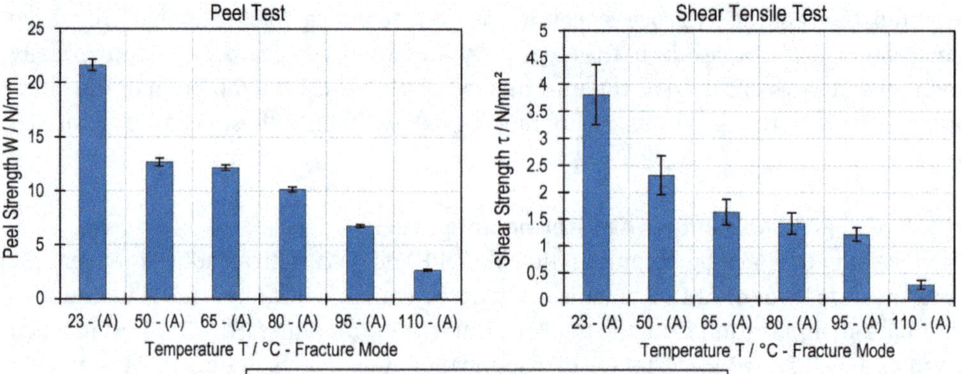

Materials		
PA12	VESTAMID NRG 2101	Evonik Industries AG
Admer	ADMER NF518E	Mitsui Chemical GmbH
HDPE	BorSafe HE3490-IS-H	Borealis AG
Test Conditions		
Testing machine: Zwick Roell Z010		
Test speed: 20 mm/min		
Test temperatures: 23, 50, 65, 80, 95, 110 °C		

Figure 4.20 Maximum peel and shear tensile strength of a composite of PA12, HDPE and an intermediate layer under the influence of test temperature

PC/ABS with TPU and EVA As Intermediate Layer

Previously, the 2-component variant of PC/ABS with TPU was already presented, showing that good adhesion exists between the two components. With increasing temperature, the peel resistance decreases continuously. At a test temperature of 95 °C, the composite can be easily separated with a peel resistance of less than 1 N/mm. With the aid of an intermediate layer of EVA, a decrease in bond strength at lower test temperatures is to be expected due to the low melting point of this material. The results of the peel test of the three-component variant of PC/ABS, EVA and TPU are shown in Figure 4.21. The EVA used is the type Escorene Ultra UL05540EH2 from the manufacturer ExxonMobil Chemical.

At room temperature, the peel resistance is somewhat lower compared to the variant without an intermediate layer, but still sufficiently high for technical applications. When the temperature is then raised to 50 °C, and thus into the range of the melting temperature of the EVA, the peel resistance drops significantly. As the test temperature is further increased, the material behavior of the EVA transitions into the plastic range, so that the resistance to peeling is attributable to the cohesive forces of the EVA and not to the adhesion forces at the interfaces.

By chemical modification of the intermediate layer, this strong drop in bond strength can be shifted to higher temperatures, so that a sufficiently high strength can be achieved under the ambient conditions typically expected in technical applications. The use of the intermediate layer demonstrates at this example, however, the possibility of establishing the recyclability of a material-bonded composite.

PP with TPU and EBA As Intermediate Layer

With regard to the material combination of PP and TPU, reference is made to the preceding discussion. By using EBA, both materials can now be brought into a material-bonded composite. The peel resistance at room temperature is somewhat lower than for the 2-component variants of PP/EBA and EBA/TPU respectively. Failure occurs between the TPU and the EBA up to a temperature of 65 °C. However, it can also be observed here that residues of the TPU remain on the EBA. At higher temperatures, the EBA partially detaches from the PP component already during heating. The bond strength decreases sharply at a temperature of 50 °C, and at 65 °C the peel resistance is already well below 1 N/mm, which can be declared as insufficient adhesion for technical applications.

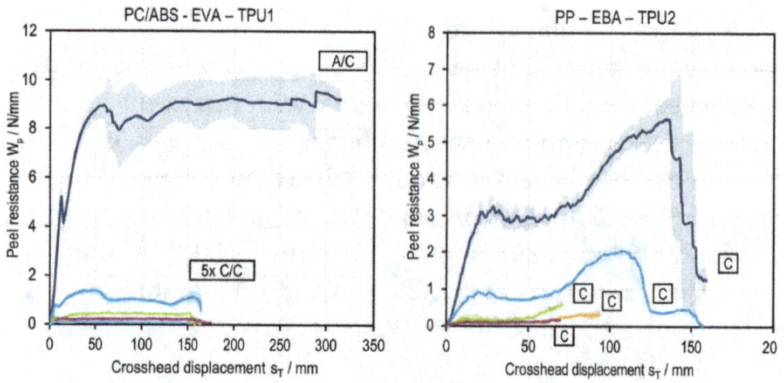

Test method: Peel test according to VDI 2019		
Materials:	Temperatures:	Fracture types:
PC/ABS T85XF	23 °C	A: no residues on component 1
EVA UL05540EH2	50 °C	B: 1-50 % residues on comp. 1
TPU 9380A (TPU1)	65 °C	
TPU 1080A (TPU2)	80 °C	C: 50-99% residues on comp. 1
PP Purell HP570M	95 °C	
EBA 1492MHG	110 °C	D: Tear-off of soft component

Figure 4.21 Peel curves for hard/soft combinations with an intermediate layer under the influence of test temperature

It can therefore be stated that the intermediate layer is what makes it possible in the first place to produce a composite of PP with TPU without surface treatment. At the same time, recyclability can be ensured, at least as far as composite separation is concerned. For this purpose, the requirements for the respective applications must be defined, further tests must be carried out and a specific material selection of the components must be made. At this point, it is merely intended to show that recyclability can already be realized with a commercially available EBA.

Summary

In Figure 4.22, the mean peel resistance normalized to the initial value at room temperature is plotted for selected material combinations at the various test temperatures. In addition, the linearly interpolated absolute value of 1 N/mm is marked on the curves, below which mechanical composite separation is assumed. While for the commercially available material combinations PC/ABS-TPU, PC/ABS-TPS and PA6GF30-TPU the mean peel resistance only falls below this value at higher temperatures, the composites in combination with the low-melting EBA show a significantly different progression.

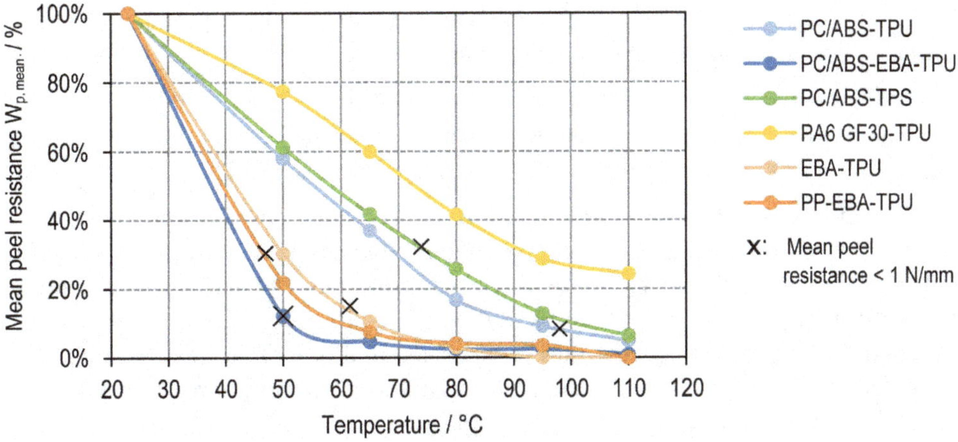

Figure 4.22 Percentage decrease in mean peel resistance for selected combinations

The drop in peel resistance is already more pronounced at temperatures of 50–65 °C and the value of 1 N/mm is reached at lower temperatures. In addition to the peel resistance shown, however, the fracture pattern must also be taken into account in each case, since cohesive failure of the bonding partners does not lead to a pure-grade separation.

Investigations into Composite Separation Using a Mixing Unit

The results presented so far are based on mechanical tests, each of which is based on a specific type of loading. In a possible recycling plant for composite separation of pre-shredded waste pieces, however, a combination of fundamental loading types occurs. Using the example of a temperature-controllable mixing unit, it is to be shown that the use of a stator-rotor system in conjunction with heated water enables composite separation of shredded waste pieces. A plowshare laboratory mixer from Gebrüder Lödige GmbH with a volume of five liters is used. Inside it, there is a plowshare centrifugal mechanism and an additional chopper head. Due to the arrangement and geometry of the plowshare blades, the mixing material is flung into the free mixing space. A suction effect, caused by the blade movement, creates a fluidized bed, which produces good mixing action in the added particulate bulk material as well as high frictional and shear forces through particle interaction. For the investigations shown, a high intensity of the centrifugal motion is achieved with a Froude number of 7. The dimensionless Froude number is a measure of the ratio of inertia to gravity and is determined according to the following functional relationship:

$$Fr = \frac{\omega^2 \times r}{g}$$

with the angular velocity ω, the largest radius r of the rotating apparatus part and the gravitational acceleration g [Pah06].

For the investigations, the following material combinations, already presented in the context of the peel investigations, are selected:

- PA6GF30 Ultramid B3EG6 with TPU Desmopan 1080A
- PP Purell HP570M with TPU Desmopan 1080A and interlayer EBA Lucofin 1492MHG

The samples produced analogously to the previously shown investigations are cut into 10 x 10 mm pieces and added to the mixer together with deionized water at a total fill level of 70 %. The volume ratio of sample material to water is varied from 1:1 to 5:1, and temperature levels of 80 °C and 95 °C are approached. The lead time for heating the mixture is approximately 10 min and the mixing time is 20 min. Afterwards, the sample pieces are counted according to separated and non-separated material. The presented results were produced within the framework of the student theses [Jäc22; Löh22].

PA6 GF30 with TPU

At a ratio of sample material to water of 2:1 and a water temperature of 80 °C, the following result is obtained after counting the sample pieces (see Table 4.2):

Table 4.2 Investigation Results Mixing Process 2-Component Composites

	PA6 G30-TPU	PA6 GF 30	TPU	Total number of composites
Number [-]	3044	7	7	3051

Less than 0.5 % of the composites can be separated. The mechanical input combined with the temperature increase is therefore not sufficient for a successful separation of the composite, which is already evident when looking at the results of the peel investigations. The TPU fulfills its function even at high temperatures by adhering well to the glass fiber-reinforced PA6.

PP with TPU and EBA As Interlayer

As shown above, the peel resistance of this material combination decreases significantly when the temperature is raised to 65 °C. This suggests that a mechanical implementation of the separation process in a bulk material consisting of particles made from the three materials used can be achieved. In this context, it must be taken into account which possible recycling pathways exist for single-grade materials or for composites that still remain intact. A remaining composite of TPU and EBA is, in the author's view, not suitable for a possible material recycling process for common TPE applications, so that a single-grade separation of the TPU should be aimed for. A blend material of PP and EBA, on the other hand, could offer advantages from a materials perspective in terms of impact strength, ductility, and flow properties. In addition, this EBA material acts as a compatibilizer for polar fillers in a polypropylene.

The results of the investigations into composite separation are presented in Figure 4.23. The left diagram shows, for the different volume ratios of sample material to process water at 80 and 95 °C, the proportion of completely separated, thus single-grade TPU particles compared to the total quantity of TPU particles. As previously described, the other process conditions such as rotational speed and fill level remain constant.

It is clearly evident that increasing the amount of sample material relative to the water up to a ratio of 3:1 leads to increased composite separation. The more sample material is contained in the water, the higher the interaction between the particles, which results in a higher mechanical load on the composite interfaces. Friction forces on the surface of the particles lead to shear forces within the particle and thus also at the interface between PP/EBA and EBA/TPU. At low ratios of sample material to water, less interaction occurs between the particles, and the shear forces transmitted by the water flow are evidently insufficient to induce composite separation.

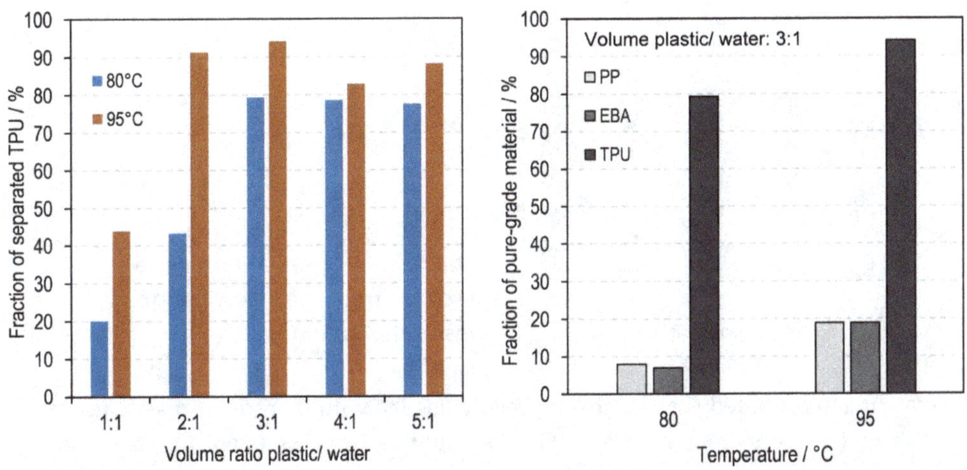

Figure 4.23 Results of the investigations into mechanical separation of the PP/EBA/TPU composite

At a temperature of 95 °C, it is observed that the proportion of separated TPU particles decreases again at ratios of 4:1 and 5:1, which contradicts the argument just presented. However, this is presumably due to the fact that the smaller amount of water at these two ratios, which is further reduced by evaporation, leads to insufficient heat transfer into the material. As a result, the composite adhesion is reduced to a lesser extent, so that the forces required to separate the composite must be higher.

The diagram on the right side in Figure 4.23 shows the single-grade fractions of the individual materials relative to the total quantity of the individual materials at a volume ratio of sample material to water of 3:1. Accordingly, with respect to the total amount of TPU in the original 3-component composite, nearly 80 % can be separated as single-grade material at 80 °C, and even over 90 % at 95 °C. In contrast, only a small fraction of less than 10 % of single-grade material is present for PP and EBA. This fraction can at least be doubled at 95 °C. This shows that the separation of the PP/EBA composite is more difficult to achieve than that of the EBA/TPU composite, which is already evident in the peel investigations.

If the evaluation is based on the scenario that a blend of PP and EBA can be reused as a recycled plastic material, the result shown in Figure 4.24 is obtained for a process temperature of 95 °C. The fractions of single-grade materials PP and EBA as well as those of the PP/EBA composites together yield a recyclable fraction of approximately 95 % at a volume ratio of 3:1, and thus similar to that of TPU as a single component.

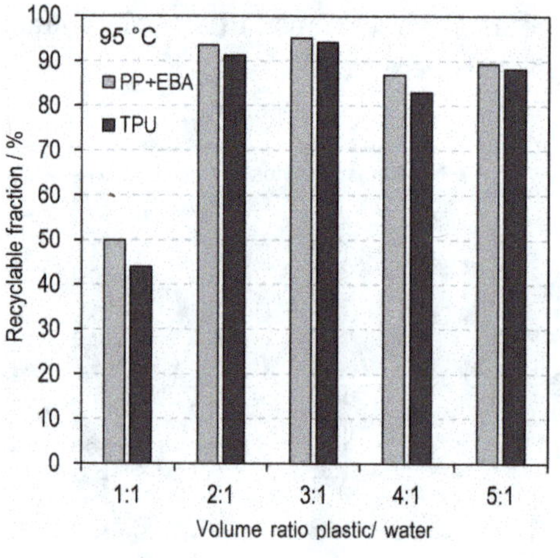

Figure 4.24 Evaluation of the recyclable fractions from the investigations into composite separation

Overall, it can be stated that for the aggregate used, a volume ratio of plastic to water of 3:1 and a process temperature of 95 °C lead to the best results with regard to composite separation. It is shown that the friction and shear forces arising from particle interaction represent the greatest effect in composite separation.

The mechanical implementation of composite separation in a stator-rotor system operated with hot water is feasible, as demonstrated using a laboratory mixing unit. Implementation for continuous operation can be realized by using a modified friction washer to which hot water is supplied. The post-sorting or separation of the individual components from the bulk material is already achievable, for example, by NIR spectroscopy.

References

[BBO+13] Baur, E.; Brinkmann, S.; Osswald, T.; Rudolph, N.; Schmachtenberg, E.: „Saechtling Kunststoff Taschenbuch". Carl Hanser Verlag, München, 2013

[BMW97] Bittner, M.; Marwick, J. von; Wolters, L.: „Aufbereitung von Kunststoffresten". In: Wolters, L.; Marvick, J. von; Regel, K.; Lackner, V.; Schäfer, B. (Hg.): „Kunststoff-Recycling. Grundlagen – Verfahren – Praxisbeispiele". Hanser-Verlag, München, 1997

[CKF+20] Chen, K.; Kroell, N.; Feil, A.; Pretz, T.: "Determination of the composition of multilayer plastic packaging with NIR spectroscopy". Multidisciplinary Journal for Waste Resources & Residues, Vol. 13, pp. 62–66, 2020

[CKW+21] Chen, K.; Kroell, N.; Wickel, J.; Feil, A.: "Determining the composition of post-consumer flexible multilayer plastic packaging with near-infrared spectroscopy". Waste Management, Vol. 123, pp. 33–41, 2021

[CMB+22] Chudalla, N.; Meschut, G.; Bartley, A.; Wibbeke, T. M.: „Bauteilschonendes Entfügen struktureller Klebverbindungen durch Kälte". In: Adhäsion Kleben + Dichten, Ausgabe 4, Springer Fachmedien Verlag, Wiesbaden 2022

[Con21] Conversio Market & Strategy GmbH: „Stoffstrombild Kunststoffe in Deutschland 2021", November 2022

[CSB+16] Cinelli, P.; Schmid, M.; Bugnicourt, E.; Coltelli, M. B.; Lazzeri, A.: "Recyclability of PET/WPI/PE Multilayer Films by Removal of Whey Protein Isolate-Based Coatings with Enzymatic Detergents". In: Materials, Vol. 9, Issue: 6, Basel, Switzerland, 2016

[DHM22] Dahlmann, R.; Haberstroh, E.; Menges, G.: „Menges Werkstoffkunde Kunststoffe", 7. vollständig neu bearbeitete Auflage, Carl Hanser Verlag, München, 2022

[Dut00] Dutschk, V.: „Oberflächenkräfte und ihr Beitrag zu Adhäsion und Haftung in glasfaserverstärkten Thermoplasten". Dissertation, Technische Hochschule Dresden, 2000

[Ehr04] Ehrenstein, G.: „Handbuch Kunststoff-Verbindungstechnik". Hanser-Verlag München, 2004

[GKH21] Gadermann, A.; Knietsch, A.; Haas, C.: „REACH: Info – REACH und Recycling". Bundesanstalt für Arbeitsschutz und Arbeitsmedizin, Dortmund, 2021

[Gri10] Grieshofer, H.: „Verbundhaftungsuntersuchungen an Mehrkomponentenspritzgussteilen". Masterarbeit, IKV – Institut für Kunststoffverarbeitung an der Montanuniversität Leoben, Begutachtung: Holzer, C., 2010

[Gut91] Gutowski, W.: "Thermodynamics of Adhesion". In: Lee, L.-H.: Fundamentals of Adhesion, Springer Verlag, Boston, MA, 1991

[Hab12] Habenicht, G.: „Kleben – erfolgreich und fehlerfrei – Handwerk, Praktiker, Ausbildung, Industrie. 6. Auflage. Vieweg+Teubner Verlag, Wiesbaden, 2012

[HHB10] Hellerich, W.; Harsch, G.; Baur, E.: „Werkstoff-Führer Kunststoffe". Carl Hanser Verlag, München, 2010

[HM21] Hopp, M; Mühlhoff, F. M.: "Separation of Multi-Component Parts for Mechanical Recycling – An Approach to Produce Hard-Soft Combinations with a Better Recyclability". In: 79th Annual Technical Conference of the Society of Plastics Engineers (ANTEC), 2021

[Hop13] Hopp, M.: „Optimierung der Verbundfestigkeit von Hart-Weichkombinationen mittels dynamischer Formnesttemperierung". Nicht veröffentlichte Masterarbeit, Kunststofftechnik Paderborn, Universität Paderborn, 2013

[HS22] Hopp, M.; Schäffer, L.: "Separation Of Multi-Component Parts For Mechanical Recycling – Study About Basic Approaches Using A Heating Process". 37th International conference of the polymer processing society (PPS), Fukuoka, Japan, 2022

[JD14] Just, D.; Diemert, J.: „Eine dufte Technik". Plastverarbeiter, Vol. 2, pp. 30–32, 2014

[JRS+11] Jepsen, Dirk; Reihlen, Antonia; Sander, Knut; Wirth, Olaf: REACH und Kunststoffrecycling – Handreichung für eine sachgerechte Umsetzung der REACH-Anforderungen für Betreiber von Recyclinganlagen. Forschungskennzahl 3708 31 305/01, im Auftrag des Umweltbundesamts, Dessau-Roßlau, September 2011

[Kai16] Kaiser, W.: „Kunststoffchemie für Ingenieure – Von der Synthese bis zur Anwendung". Carl Hanser Verlag, München, 2016

[Käu87] Käufer, H.: „Verfahren zum Entkleben von Metall-Nichtmetall-Klebungen". Deutsches Patent, DE3823952C2, 1987

[Kle11] Kleeschulte, R.: „Modellierung der Verbundfestigkeit von Hart-Weich-Kombinationen polymerer Werkstoffe". Dissertation, Kunststofftechnik Paderborn, Universität Paderborn, 2011

[Kuh98] Kuhmann, K.: „Prozess- und Materialeinflüsse beim Mehrkomponentenspritzgießen". Dissertation, Friedrich-Alexander-Universität Erlangen-Nürnberg, 1998

[Küh05] Kühnert, I.: „Grenzflächen beim Mehrkunststoffspritzgießen". Dissertation, Technische Universität Chemnitz, 2005

[LKS17] Lieberwirth, H.; Krampitz, T.; Schnutz, S.: „Aufschlusszerkleinerung metallintensiver Leichtbaustrukturen zur Freilegung von Wertstoffkomponenten am Beispiel einer Hybrid-B-Säule in Mischbauweise". In: Recycling und Rohstoffe, Band 10, pp. 327–337, Hrsg: Thomè-Kozmiensky, K.; Thiel, S., TK Verlag, Neuruppin, 2017

[MDK+20] Moritzer, E.; Deuse, J.; Kroll, A.; Richter, R.; Hopp, M.; Schmitt, J.; Schulte, L.; Schrodt, A.; Wittke, M.: „Einsatz von maschinellem Lernen für die Rezyklat-Verarbeitung". In: Jahresmagazin Kunststofftechnik, pp. 70–73, 2020

[MG16] Martens, H.; Goldmann, G.: „Recyclingtechnik", Springer Fachmedien, Wiesbaden, 2016

[MHK13] Moritzer, E.; Hopp, M.; Kleeschulte, R.: "Improving bond strength of hard-soft combinations by dynamic mold heating". In: TPE magazine international, Dr. Gupta Verlag, Vol. 5, pp. 176–183, Ratingen, 2013

[MK03] Miller-Chou, B. A.; Koenig, J. L.: "A review of polymer dissolution". In: Progress in Polymer Science, Vol. 28, pp. 1223–1270, 2003

[Nik05] Nikolova, D.: „Charakterisierung und Modifizierung der Grenzflächen im Polymer-Metall-Verbund". Dissertation, Martin-Luther-Universität Halle-Wittenberg, 2005

[NL94] Nauman, E. B.; Lynch, J. C.: "Polymer recycling by selective dissolution". US5278282A, US-Patent, 1994

[NN01-ol] Deutsches Verpackungsinstitut e. V., 2022: Verbundmaterial Link: https://p239009.mittwaldserver.info/materialien.html, zuletzt abgerufen am 28.04.2022

[NN02-ol] APK AG: https://www.apk-ag.de/newcycling/prozess/, zuletzt abgerufen am 27.09.2022

[NN03-ol] APK AG: https://www.apk-ag.de/apk-stellt-vollentfaerbte-kunststoff-rezyklate-vor/, zuletzt abgerufen am 27.09.2022

[NN04-ol] Saperatec: https://www.saperatec.de/de/technologie.html, zuletzt abgerufen am 27.09.2022

[NN05-ol] Kunststoffweb: „Saperatec: Recyclinganlage für Verbundfolien in Dessau". Pressemitteilung vom 01.09.2022, Link: https://www.kunststoffweb.de/branchen-news/saperatec_recyclinganlage_fuer_verbundfolien_in_des-sau_t251063, zuletzt abgerufen am 27.09.2022

[NN06-ol] Gebrüder Lödige Maschinenbau GmbH; 2022; Pflugschar – Mischer von Lö-dige (Labormischen) Link: https://www.loedige.de/de/maschinen/labor-pflug-scharmischer/#&gid=1&pid=3, Zugriff: 19.08.2022

[NN07-ol] GVM Gesellschaft für Verpackungsmarktforschung: „Aufkommen und Verwertung von PET-Getränkeflaschen in Deutschland 2021". Link: https://newsroom.kunststoffverpackungen.de/2023/03/28/einsatz-von-rpet-in-pet-getraenkeflaschen-steigt-auf-448-prozent/, zuletzt abgerufen am 14.07.2023

[NN08-ol] Stiftung Zentrale Stelle Verpackungsregister: „Mindeststandard für die Bemessung der Recyclingfähigkeit von systembeteiligungspflichtigen Verpackungen gemäß § 21 Abs. 3 VerpackG", 2019 Link: https://www.Verpackungsregister.org/fileadmin/files/Mindeststandard/Mindeststandard_VerpackG_2021.pdf, zuletzt abgerufen am 28.12.2022

[NN09-ol] Veolia: Bottle-to-bottle-Recycling, Link: https://www.veolia.de/bottle-to-bottle-recycling, zuletzt abgerufen am 28.12.2022

[NN10-ol] Kunststoffweb: „Recyclingfähigkeit: PE-Folien mit EVOH-Barriereschichten im Test". Pressemitteilung vom 11.11.2019, Link: https://www.kunststoffweb.de/branchen-news/recyclingfaehigkeit_pe-folien_mit_evoh-barriereschichten_im_test_t243864, zuletzt abgerufen am 29.12.2022

[NN11-ol] EREMA Engineering Recycling Maschinen und Anlagen Ges.m.b.H.: "Powerful Filtration", Link: https://www.erema.com/de/filtrieren/, zuletzt abgerufen am 14.01.2023

[NN12-ol] Nordson BKG GmbH: Technisches Datenblatt „BKG® HiCon™ R-Type 250", Link: https://nc-p-001.sitecorecontenthub.cloud/api/public/content/6c294d57e7e342c0b538274561cedc51?v=ca107871, zuletzt abgerufen am 14.01.2023

[NN13-ol] EREMA Engineering Recycling Maschinen und Anlagen Ges.m.b.H.: "ReFresher", Link: https://www.erema.com/de/refresher/, zuletzt abgerufen am 24.01.2023

[Nor18] Nordmeyer, T.: „Verfahrenstechnische Entwicklung des Direktinjektion-Plasmaverfahrens im Spritzgießprozess". Dissertation, Kunststofftechnik Paderborn, Universität Paderborn, 2018

[Pah06] Pahl, M.: „Mischer und deren Mischungsprinzipien", Vorlesungsskript, Institut für Energie- und Verfahrenstechnik, Universität Paderborn, 2006

[Pot04] Potente, H.: „Fügen von Kunststoffen – Grundlagen, Verfahren, Anwendung". Carl Hanser Verlag, München, 2004

[PSW+11] Plog, S.; Schneider, J.; Walker, M.; Schulz, A.; Stroth, U.: "Investigations of plasma polymerized SiOx barrier films for polymer food packaging". In: Surface and Coatings Technology, Vol. 205, Supplement 2, pp. S165–S170, https://doi.org/10.1016/j.surfcoat.2011.01.034, 2011

[Ras12] Rasche, M.: „Handbuch Klebtechnik“. Carl Hanser Verlag, München, 2012

[RHB12] Rösler, J.; Harders, H.; Bäker, M.: „Mechanisches Verhalten der Werkstoffe“. Springer Vieweg, Wiesbaden, 2012

[Rot09] Rotheiser, J.: "Joining of Plastics". Carl Hanser Verlag, München, 2009

[Sch02] Schubert, G.: „Zerkleinerungstechnik für nicht-spröde Abfälle und Schrotte“. In: Aufbereitungstechnik, Vol. 43 (9), pp. 6–23, 2002

[Sch20] Schöppner, V.: „Standardverfahren Extrusion“. Vorlesungsskript, Kunststofftechnik Paderborn, Universität Paderborn, 2020

[Sch22] Schöppner, V.: „Fügen von Kunststoffen“. Vorlesungsskript, Kunststofftechnik Paderborn, Universität Paderborn, 2022

[SFM+20] Schlummer, M.; Fels, T.; Mäurer, A.; Altnau, G.: „Die Rolle der Chemie beim Recycling“. In: Kunststoffe, 06/2020, Carl Hanser Verlag, München, 2020

[SNK+22] Schlummer, M.; Neubauer, J.; Knappich, F.; Mäurer, A.; Weig, U.; Backfisch, M.; Rinderhofer, A.: „Recycling von Verbundfolien“. In: Kunststoffe, 01/2022, Carl Hanser Verlag, München, 2022

[SS22-ol] Schüssler, A.; Schneider, H.: „Desodorierung von recyceltem Kunststoff“. Prozesstechnik online, Oktober 2022

[TM06] Tucker, J. D.; Myers, D. L.: Separation Process for Multi-Component Polymeric Materials. US-Patent, US 7087655 B2, 2006

[Ueb68] Ueberreiter, K.: "The solution process". In: Crank, J.; Park, G. S.: Diffusion in polymers. New York, NY: Academic Press, pp. 219–257, 1968

[VP91] Vakula, V. L.; Pritykin, L. M.: "Polymer Adhesion – basic physico-chemical principles". Ellis Horwood Limited, Chichester, West Sussex, England, 1991

[Woi14] Woiton, M.: „Netzartig strukturierte Oberflächen aus präkeramischen Polymeren“. Dissertation, Friedrich-Alexander-Universität Erlangen-Nürnberg, 2014

[ZLN18] Zhao, Y.; LV, X.; Ni, H.: "Solvent-based separation and recycling of waste plastics: A review". In: Chemosphere, Vol. 209, pp. 707–720, https://doi.org/10.1016/j.chemosphere.2018.06.095, 2018

[ZSL21] Zepnik, S.; Simnick, A.; Lauter, M.: „Wesentliche Aspekte beim 2 K-Spritzgießen mit TPU als Weichkomponente“. In: GAK Gummi Fasern Kunststoffe, Jahrgang 74, Vol. 10, Dr. Gupta Verlag, 2021

Student theses

The student theses listed below were prepared in the context of the present work at the Chair of Plastics Technology at the University of Paderborn. The definition of objectives, the execution as well as the evaluation, interpretation and visualization of results were carried out under the scientific supervision of the advisor Dr.-Ing. Matthias Hopp. The results obtained have in part been incorporated into this work and into the publications cited in the text.

[Fie21] Fiene, E. (Hopp, M.; Moritzer, E.): „Haftungsuntersuchungen an Hart-Weich-Kombinationen aus dem 2 K-Spritzgießprozess unter Temperaturbeanspruchung zur Bewertung der Recyclingfähigkeit". Universität Paderborn, Kunststofftechnik Paderborn, unveröffentlichte Studienarbeit Juli 2021

[Han22] Hanselle, F. (Hopp, M.; Schöppner, V.): „Ermittlung einer geeigneten Temperaturführung und Schneckenkombination zur Verbundtrennung von thermoplastischen Kunststoffverpackungen im gleichläufigen Doppelschneckenextruder". Universität Paderborn, Kunststofftechnik Paderborn, unveröffentlichte Studienarbeit April 2022

[Jäc22] Jäckel, A. K. (Hopp, M.; Moritzer, E.): „Experimentelle Untersuchungen zur Verbundtrennung von 2 K- und 3 K-Kunststoffverbunden im Hinblick auf das werkstoffliche Recycling". Universität Paderborn, Kunststofftechnik Paderborn, unveröffentlichte Masterarbeit September 2022

[Kir22] Kirstein, S. (Hopp, M.; Moritzer, E.): „Verbundtrennung von Mehrschichtverpackungen". Universität Paderborn, Kunststofftechnik Paderborn, unveröffentlichte Bachelorarbeit Januar 2022

[Löh22] Löhr, F. (Hopp, M.; Moritzer, E.; Schäffer, L.): „Untersuchung zur Verbundtrennung von 2 K- und 3 K-Kunststoffbauteilen". Universität Paderborn, Kunststofftechnik Paderborn, unveröffentlichte Projektarbeit Juni 2022

[Lüt22] Lütkebohle, P. J. (Hopp, M; Moritzer, E.): „Untersuchungen von spritzgegossenen 2 K-Probekörpern in Bezug auf Verbundhaftung, Zerkleinerung und Einrieselverhalten in einen gleichläufigen Doppelschneckenextruder". Universität Paderborn, Kunststofftechnik Paderborn, unveröffentlichte Studienarbeit April 2022

[Rem20] Rempel, T. (Hopp, M; Moritzer, E.): „Grundlegende Untersuchungen zur Recyclingfähigkeit von Hart-Weich-Kombinationen aus dem 2 K-Spritzgießprozess". Universität Paderborn, Kunststofftechnik Paderborn, unveröffentlichte Bachelorarbeit September 2020

[Sch20] Scholz, C. (Hopp, M.; Moritzer, E.): „Analyse der Bedeutung des Kunststoffrecyclings für die Kunststoffbranche unter Berücksichtigung aktueller, gesellschaftspolitischer Entwicklungen". Universität Paderborn, Kunststofftechnik Paderborn, unveröffentlichte Masterarbeit Juli 2020

[Wes22] Westphal, M. S. (Hopp, M.; Moritzer, E.): „Untersuchung der Verbundtrennung durch Temperatureinfluss von Hart-Hart-Kunststoffverbunden für ein sortenreines, werkstoffliches Recycling". Universität Paderborn, Kunststofftechnik Paderborn, unveröffentlichte Bachelorarbeit Februar 2022

5 Feedstock Recycling

With only 0.5%, feedstock recycling in Germany in 2021 accounts for an extremely small share of the total utilization of plastic waste. While at the beginning of the 2000s the amount of waste fed into feedstock recycling was above 300,000 tonnes for several years, it has decreased significantly in recent years and most recently fell by 70% from 2017 to 2019 to 14,000 tonnes [Con19, p. 99]. In 2021, this amount increased slightly again to 26,000 tonnes. Feedstock recycling is expected to play an important role in the future, as documented by the major financial efforts of the petrochemical industry. The advantages and disadvantages of feedstock recycling compared to material recycling will be briefly discussed in the next section. The subsequent sections describe the chemical processes that lead to the conversion of polymers into basic chemical feedstocks.

5.1 Overview and Definitions

In feedstock recycling, the polymers are converted into monomers, basic chemical building blocks or pyrolysis gas via chemical processes. This occurs under the influence of heat, catalysts, solvents or hydrogen. Thus, intentional material degradation, an intentional shortening of the polymer chains, takes place. In contrast to material recycling, it is not the material plastic that is recovered here, but rather the raw materials (monomer, oil or gas). Since these are chemical processes, feedstock recycling is frequently referred to as chemical recycling.

The advantage of this type of recycling over material recycling is that heavily mixed and contaminated waste can also be used, although a certain degree of purity must still be present depending on the process. This would allow material cycles to be closed for waste that has previously been used for energy recovery. A further advantage over material recycling is that through the possible recovery of monomers, new plastic can be polymerized which exhibits virgin material quality. The disadvantage of chemical recycling processes lies, among other things, in the high operating costs and the difficulty of scalability.

Figure 5.1 shows the cycle of plastics for chemical recycling. After product manufacturing and sales, the plastic component or plastic packaging transitions into the use phase. Disposal forms the transition to the processing phase, which in the case of chemical recycling falls into the area of Waste to Polymers and Waste to Chemicals can be divided. The former shows the cycle for plastics, in which monomers are produced through chemical processes that can then be polymerized again. The resulting plastic has virgin material quality and can be used for new plastic products. Mixing in virgin material, as is necessary when using recyclate from material recycling, is not required.

In the area of Waste to Chemicals waxes, oils or gases for the petrochemical industry are produced through conversion processes of the plastics. These raw materials are thus withdrawn from the plastics cycle, but are available for the production of chemicals. The sorting and shredding process precedes all chemical processes. Shredding is necessary, among other things, to increase the surface area so that solvents and other auxiliaries can completely and comprehensively wet the waste pieces [MG16, S. 309].

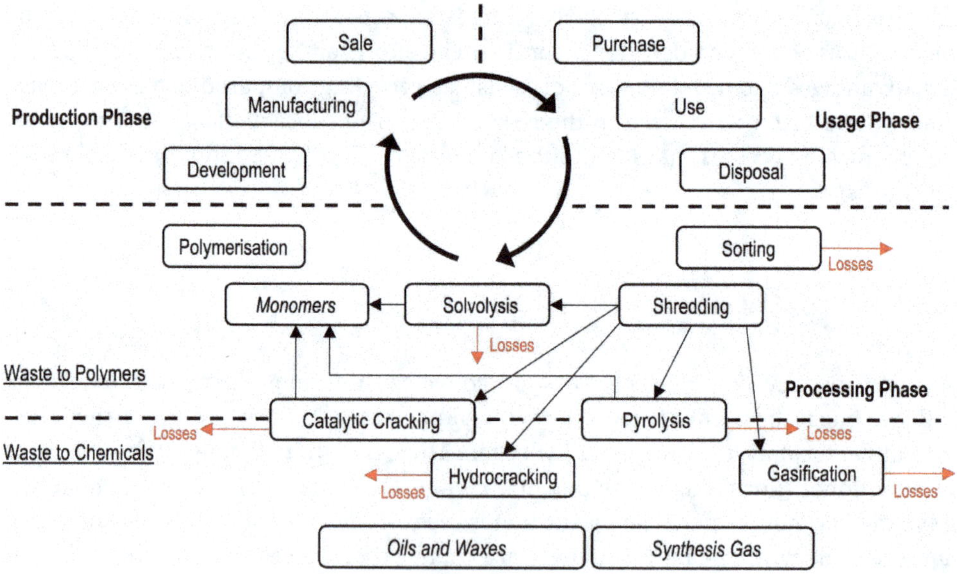

Figure 5.1 Overview of the circular economy processes in feedstock recycling

The processes of feedstock recycling examined in more detail in the following sections can be classified according to [LST+20] distinguish in

- Solvolysis (chemical depolymerization
- Pyrolysis (thermal conversion)
- Catalytic cracking (thermocatalytic conversion)
- Hydrocracking
- Gasification or gasification

From solvolysis, pyrolysis and partly also from catalytic cracking, monomers can be obtained. In addition, oils, waxes or gases are produced from all processes.

5.2 Solvolysis

In solvolysis, sometimes also called chemolysis or chemical depolymerization, plastics are broken down into their basic building blocks (monomers) with the aid of solvents and, if necessary, additional chemicals. These are considered qualitatively equivalent to the primarily produced monomers and can be used again for the synthesis of new plastics. Solvolysis represents the reverse reaction of step-growth reactions (condensation polymerization and addition polymerization), whereby it is mainly used for polycondensates. It is therefore particularly suitable for PET, PA or PU, which together, however, account for only 15% of the total plastic waste volume in Germany [Con21, p. 88]. In addition, a large proportion of PET can be recycled as material, which represents an established treatment pathway particularly in the area of PET bottles [CAS99, p. 31; LST+20, p. 50].

The plastic waste is first shredded for solvolysis and foreign substances are, as far as possible, separated out (see Figure 5.2). The attack of the solvent occurs at the surface of the waste pieces, which is why a waste stream shredded and homogenized to a size of 10 mm is required in order for the solvolysis reaction to proceed efficiently [MG16, p. 309].

The solvolysis can be further subdivided depending on the type of solvent into

- Hydrolysis
- Glycolysis
- Methanolysis
- Ammonolysis

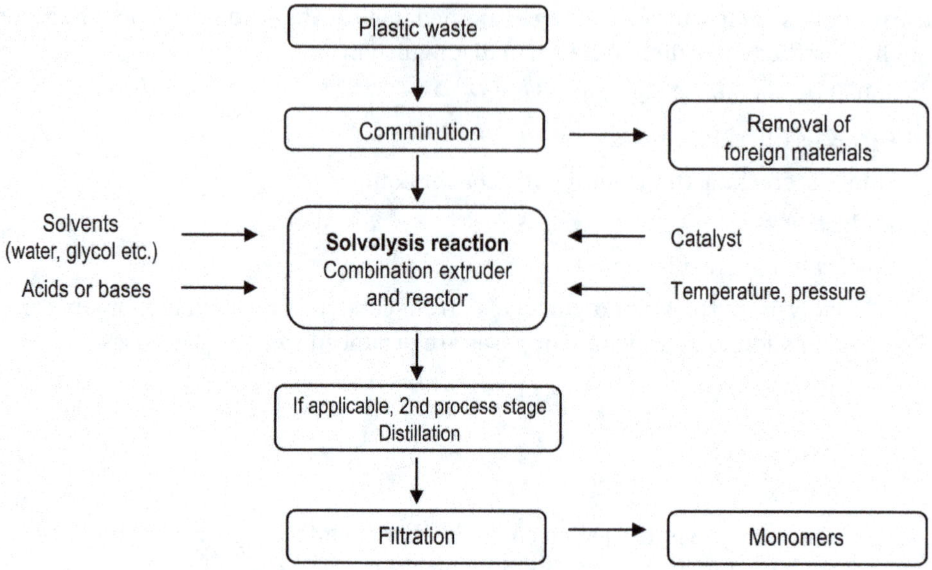

Figure 5.2 Overview of the process steps in solvolysis

The processes in which alcohols are used as solvents are also collectively referred to as alcoholysis, while in hydrolysis water is used as the solvent [CAS99, p. 32, MG16, p. 309].

In addition to the solvent, acids or bases and sometimes a catalyst are added to the process, which takes place under pressure and at a defined temperature in a reactor or in combination with an extruder. This is followed by filtration, purification and distillation steps, which are used depending on the substance to be recovered. In this way, dyes and other interfering substances can be removed from the mixture of solvent and polymer/monomer, and the monomer can finally be precipitated.

The solvolysis processes are thus suitable for contaminated PA, PU, or PET waste streams. Composite materials can also be broken down by dissolving only the plastic that is to be separated into its monomers, while the plastic present in the material composite is removed by filtration. However, this process must be distinguished from selective dissolution in the course of material recycling, because in solvolysis the polymer structure of the dissolved plastic is not preserved, but is instead deliberately degraded after or during dissolution.

According to a compilation of current commercialization approaches for the chemical recycling of [LST+20] the company DuPont Teijin Films operates the LuxCR™ process for the solvolysis of PET for its own further use. The company Eastman is planning a methanolysis plant for PET in France with a capacity of 160,000 tonnes [NN01-ol; Göb22-ol]. Further processes and plants are frequently still at laboratory scale.

Solvolysis of PET

For over 50 years, the glycolysis of PET has been known, in which PET is broken down into the chemical compound bis(hydroxyethyl)terephthalate (BHET) and further oligomers using glycol. The process takes place at temperatures between 180 and 240 °C and can be accelerated by the addition of catalysts. The latter can also be achieved through the use of microwaves. The concentration of the catalyst, the reaction time and the reaction temperature thus have a great influence on the process. The resulting compounds can be used for the synthesis of polyurethanes, copolyesters, unsaturated resins or for the renewed synthesis of PET [SP19; LST+20; RGD17].

Figure 5.3 shows the further solvolysis processes for PET with an overview of the resulting substances. In methanolysis, methanol is used as a solvent, which diffuses into the polymer structure at temperatures between 180 and 280 °C and a pressure between 20 and 40 bar. A catalyst is also frequently used. The substances resulting from chain cleavage are dimethyl terephthalate and ethylene glycol, which can be used for PET synthesis. Higher purities can be achieved through methanolysis than with glycolysis, however higher purities are also necessary for the input waste stream. A two-stage process frequently takes place, in which a glycolysis to diglycol terephthalate is first carried out and then a methanolysis to DMT. In this way, a PET can be synthesised which is also approved for food contact [CAS99, p. 38; LST+20; p. 51; MG16, p. 311 f.].

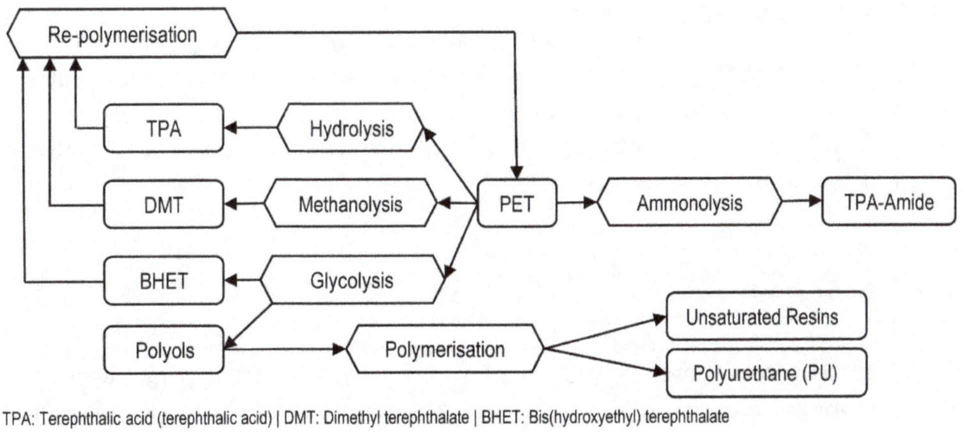

TPA: Terephthalic acid (terephthalic acid) | DMT: Dimethyl terephthalate | BHET: Bis(hydroxyethyl) terephthalate

Figure 5.3 Solvolysis routes for PET according to [CAS99]

In hydrolysis of PET, terephthalic acid and ethylene glycol are produced, albeit in such low purities that this process is of lesser importance compared to alcoholysis. From ammonolysis, diamides of terephthalic acid can be obtained, but these are difficult to convert for PET synthesis [CAS99, p. 38–41].

The machine manufacturer KraussMaffei from Hanover offers a process for PET recycling in which polymer degradation via glycolysis does not occur completely, but only to the extent that the melt viscosity drops significantly. The PET waste stream enters a co-rotating twin-screw extruder, where it is melted, homogenized and freed from moisture and other volatile components by means of vacuum degassing. Liquid glycol is then injected, which leads to chain scission and the formation of oligomers. By reducing the viscosity, filters with high fineness (<60 µm) can be used. The purified, low-viscosity melt is subsequently repolymerized in a reactor [KP22].

Solvolysis of PA

For polyamide waste, the hydrolysis process is used. Under pressure and elevated temperature, with the action of acids or bases, a catalyst and the solvent water, the PA is cleaved into the monomer caprolactam. After the hydrolysis reaction, filtration and distillation are carried out to obtain a purified product that can be further processed. Caprolactam can then be polymerized again to form new polyamide [MG16, p. 309 f.].

Solvolysis of PU

In the chemical depolymerization of polyurethane, water or glycol are used as solvents. Hydrolysis produces polyols and amines, the latter of which cannot be directly utilized. Furthermore, the processes have not yet been economically viable and, according to current knowledge, only pilot plants exist [ZPV+21, p. 3]. Processes with glycol are complex and require auxiliary agents that prevent the formation of aromatic amines. Nevertheless, this process is state of the art for PU components as well as for PU foams. In addition, the polyol can be used for the renewed production of PU [MG16, p. 310; ZBB07, p. 683 f.]. Through a multi-stage process consisting of glycolysis and hydrolysis, in addition to polyol, amines for the renewed production of isocyanate from PU can also be recovered. Recently published research results demonstrate the technical potential of this approach [ZPV+21].

5.3 Pyrolysis

Pyrolysis is a process known for centuries for thermally decomposing organic materials in the absence of oxygen. Plastics can also be broken down by pyrolysis into shorter hydrocarbon chains, which can be further processed in various ways. The numerous process variants that have been developed are as diverse as the substances obtained from them. Reaction products are formed that can be used to produce monomers or that serve as waxes or oils as feedstocks for the petrochemical industry and for fuels.

As with all chemical recycling processes, the comminution of plastic waste for further processing is essential (see Figure 5.4). The presence of contaminants is generally not a major problem for the process, as these decompose completely. The pyrolysis reaction takes place in a reactor under exclusion of air or oxygen. A distinction is made between slow pyrolysis (below 400 °C), moderate (400–600 °C) and fast pyrolysis (above 600 °C).

In addition to the necessary residence time for complete conversion of the starting materials, the temperature influences the type of products formed. Higher temperatures lead to the formation of light products, gases and coke, while the formation of waxes and heavy oils occurs at lower temperatures [LST+20; QNN+17, p. 17]. Following the pyrolysis process, distillation and refining steps are carried out depending on the substances to be recovered [MG16, p. 316 f.; LST+20, p. 52 ff.; RDG17, p. 21; SKP20, p. 107].

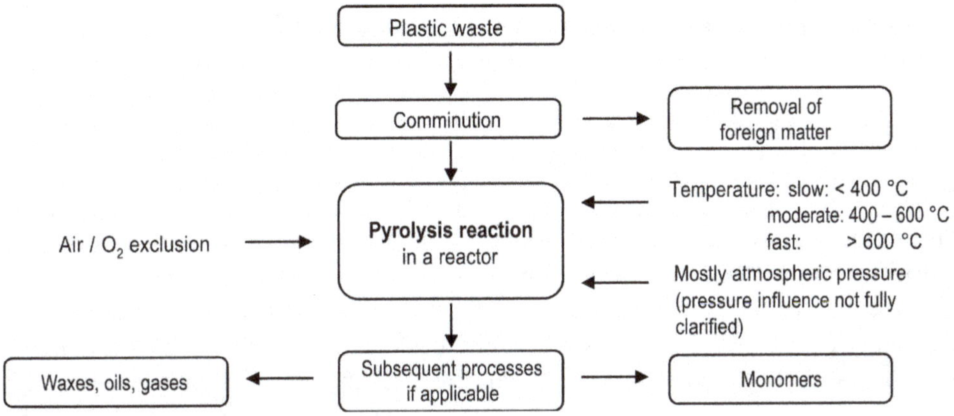

Figure 5.4 Overview of the process steps in pyrolysis

Depending on the reactor type, different requirements are placed on the input stream with regard to homogeneity, dust content or feedability. Furthermore, care must be taken to ensure that certain plastics are not present in the waste stream. PVC in particular must be mentioned here, as it forms hydrogen chloride during pyrolysis and can cause corrosion. In fluidized bed reactors, the hydrogen chloride can be bound, for example by adding lime, if contamination with PVC cannot be avoided [Kam93]. With PET, the formation of terephthalic acid can occur, which precipitates and then leads to blockages in the plant [LST+20]. The literature frequently describes processes for the pyrolysis of PU in which, for example, ethylene or methane are produced as products [FFC+01; ZBB07]. The use of the polyolefins PE and PP for pyrolysis reactions is well suited, as only carbon and hydrogen occur as elements. PA, PS and PMMA can be converted to a high degree into their monomers by pyrolysis with subsequent process steps [RDG17; KPS04].

According to research conducted by the Wuppertal Institute in 2020, of 133 pyrolysis processes published in the literature, only two are at pilot scale. The other 131 processes are still being tested in laboratories. The input stream predominantly consists not of mixed waste but of pure materials. In industry, numerous projects are planned to build pyrolysis plants on a larger scale. The company Agilyx claims to operate the world's first commercial pyrolysis plant for the feedstock recycling of polystyrene. In this process, the monomer styrene is recovered. The process used has also been sold to the polystyrene manufacturer INEOS Styrolution [RVM+20, p. 17; Tul19-ol].

Many pyrolysis processes are complex in their design and maintenance-intensive; furthermore, a large amount of energy is required for operation. The input stream must be prepared and continuously fed for trouble-free operation. Scaling up to a commercial plant size has so far been an insurmountable hurdle for many of the processes. The recent efforts of the major chemical industry to operate these processes economically within the framework of joint ventures with partner companies could lead to an increase in the share of feedstock recycling in the future.

5.4 Catalytic Cracking

Catalytic cracking is carried out as a process similar to pyrolysis. The essential difference lies in the use of a catalyst. Thermal cracking takes place at temperatures between 300 and 900 °C in the absence of oxygen (see Figure 5.5). Heterogeneous systems such as zeolites, silica, alumina, metal oxides or metal-coated carbon are most commonly used as catalysts. Less frequently used homogeneous systems are based on Lewis acids (e.g. aluminum trichloride). The resulting products are oils and gases, which can be used, for example, for alternative fuels or for basic chemicals [ASE08; LST+20; MBA+16].

The catalyst reduces the activation energy for the cracking process, which allows the temperature and reaction time to be kept lower. In addition, the choice of catalyst allows better control of product yield than in pyrolysis. More cyclic, branched and aromatic compounds are formed, which offer more possibilities in product utilization than is the case with pyrolysis [AM16; LST+20; MBA+16].

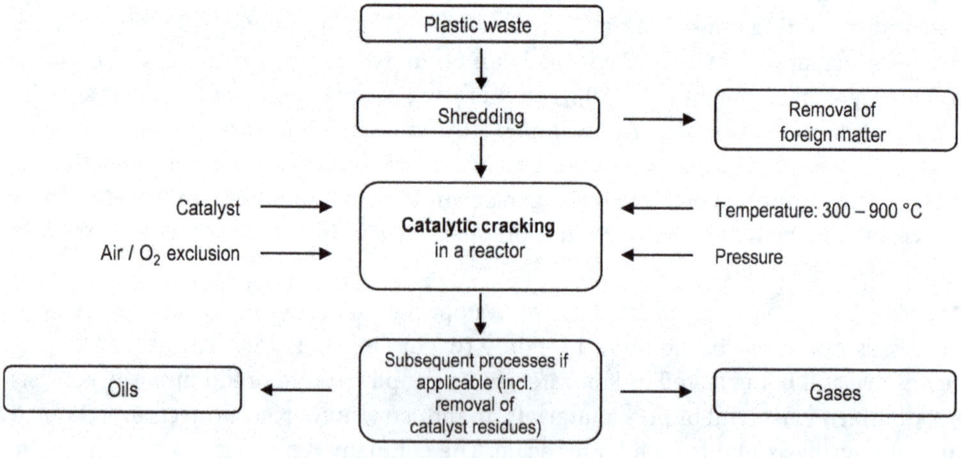

Figure 5.5 Overview of the process steps in catalytic cracking

When using the catalyst, care must be taken to ensure that the waste stream does not contain any substances that could neutralize the catalyst's effectiveness. These can include, for example, heteroatoms such as sulfur or nitrogen, which enter the process via certain types of plastics. The presence of mineral fillers or coke formed during the process are also possibilities that can impair the catalyst's effectiveness. By employing additional equipment, interfering substances can be removed from the process [AM16; LST+20].

Overall, catalytic cracking also offers a wide variety of possible process engineering configurations depending on the input stream. For real waste streams, the analysis of potential catalyst poisons is important. In addition, the connection to downstream processing steps is decisive for the economical operation of the plant. Current industrial initiatives in the field of catalytic cracking include, for example, MoReTec by LyondellBasell [NN02-ol], Quantafuel [NN03-ol], or Cassandra Oil [NN04-ol].

5.5 Hydrocracking

By adding hydrogen, hydrocracking can produce more stable hydrocarbon compounds and thus higher product qualities. Typical products are, for example, naphtha or diesel [RDG17, p. 24; LST+20].

The process frequently takes place in two stages, with the first stage involving pyrolysis with or without a catalyst at low temperature. In the second stage, hydrogen is added to the process under a partial pressure of 200–250 bar, and saturated hydrocarbons are formed (see Figure 5.6). The advantage of this variant is that between the pyrolysis and the hydrogenation step, interfering substances, heteroatoms, or coke can be removed that might potentially harm the catalyst in the hydrogenation step. Bifunctional catalysts are used in hydrocracking, which on the one hand support the cracking of the polymer chains and on the other hand have a hydrogenating function [LST+20].

The energy requirement is, as with catalytic cracking, lower than with pyrolysis due to the use of the catalyst. Since hydrogenation is an exothermic reaction, the heat generated can additionally be used for chain shortening. The high costs for the catalyst and hydrogen are disadvantageous. For this reason, many plants had to be shut down again in the past [MG16, p. 314; LST+20, p. 56]. The energy recovery competing with mixed plastic waste (see Chapter 6) has clear economic advantages over this process.

The patented and multiply award-winning two-stage Handerek process of the company Handerek Technologies sp. z.o.o. from Poland has, since the company's founding in January 2018, represented an attempt to industrialize the process.

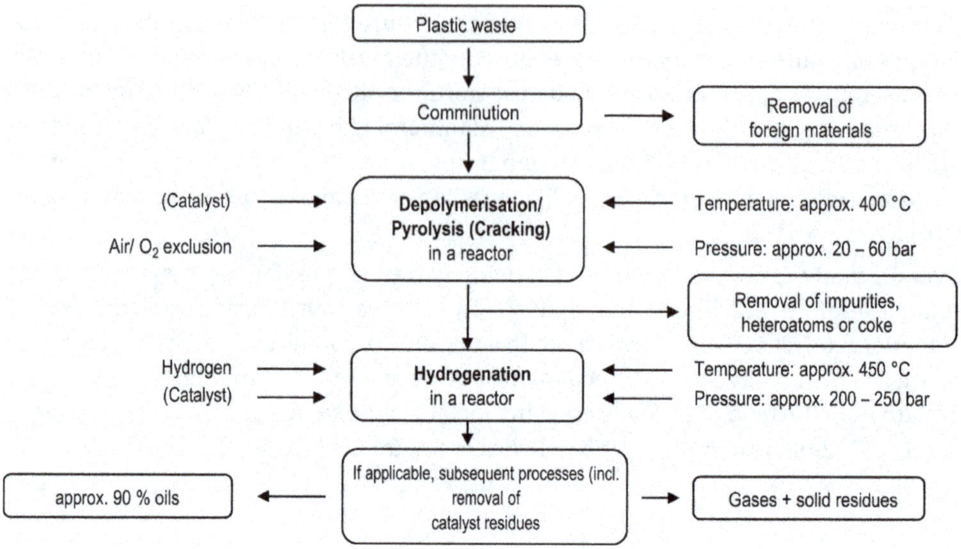

Figure 5.6 Overview of the process steps in hydrocracking

According to company information, contaminated plastic waste from PP, PE, and PS can be used as the input stream. In the first stage, thermocatalytic cracking takes place. The resulting vapor condenses and is fed to the hydrogenation step. After fractionation, naphtha and diesel can be withdrawn as products [NN05-ol].

5.6 Gasification

While pyrolysis is carried out under exclusion of oxygen, gasification is achieved by the partial oxidation of carbon under pressure and high temperature. The gasification of coal to produce synthesis gas is a long-established process. In addition, plastic waste can also be used for gasification. When the resulting synthesis gas is used for the production of basic chemicals, the process is classified as feedstock recycling. Alternatively, the gas can also be used as an energy carrier [MG16, p. 317; VKF20, p. 12].

No high demands are placed on the plastic waste used with regard to purity and presorting, as is the case with pyrolysis. Rather, the plastics are fed into the gasification process together with other solid materials such as biogenic residues or municipal solid waste [LST+20, p. 56]. Oxygen, steam, carbon dioxide, or in rare cases hydrogen are used as gasification agents. The resulting synthesis gas contains varying proportions of carbon monoxide (CO), hydrogen (H_2), and methane (CH_4). Several utilization pathways are available for the synthesis gas, for example the production of methanol or ammonia. The partially catalytically operated processes required for this demand a gas purification of the synthesis gas. In particular, nitrogen- or chlorine-containing components must be removed. Chlorine and sulfur can already be bound in the reactor by the addition of lime, as is the case, for example, with the process of ecoloop GmbH [LST+20; RDG17, p. 27 Möl12].

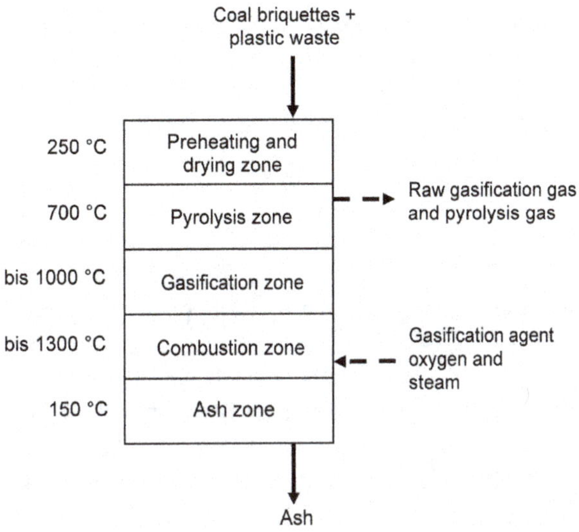

Figure 5.7 The gasification process using the example of the fixed-bed pressure reactor, after [MG16, p. 318]

Figure 5.7 schematically shows the design of a fixed-bed pressure reactor for the gasification of plastic waste at a temperature of up to 1300 °C and a pressure of 25 bar. This reactor operates on the countercurrent principle, meaning that the gasification agent is introduced from below while the waste is fed in from above. The raw gases, in turn, are discharged at the top, and the combustion residues are discharged at the bottom as ash. The bulk material must be distributed as uniformly as possible and ensure good gas flow through it for complete combustion. For this purpose, the waste plastics and the other feed materials such as coal are previously briquetted or agglomerated. During operation, the reactor exhibits different zones that differ in particular in temperature. At the very top is the preheating and drying zone, followed by the pyrolysis zone, in which thermal decomposition into short-chain hydrocarbon compounds takes place. This is followed by the actual gasification zone at a temperature of up to 1000 °C, in which endothermic reactions lead to the formation of carbon monoxide (CO) or methane (CH_4). Higher temperatures lead to increased formation of CO and H_2, while higher pressure results in higher proportions of CH_4. In the combustion zone, in which, among other things, the exothermic reaction $C + O_2 \rightarrow CO_2$ takes place, the highest temperatures of up to 1300 °C are present. The energy released in this zone can be used in the gasification zone for the endothermic reactions [LST+20, p. 56; MG16, p. 317 f.].

A study by the Wuppertal Institute has shown that, based on the current state of research, ten processes for gasification with a technology readiness level of > 6 are opposed by a further 41 processes that exist only at laboratory scale. This shows that gasification processes have a higher maturity level than pyrolysis processes; however, here too there are only a few plants at a large scale. As an example of a large-scale plant, the Enerkem process of the Canadian company Enerkem with a capacity of 360,000 t/a should be mentioned, which uses municipal solid waste as input stream and produces ethanol and methanol as output [LST+20, p. 58; NN06-ol; RVM+20, p. 16].

References

[AM16] Almeida, D.; Marques, M. d. F.: "Thermal and catalytic pyrolysis of plastic waste". Polimeros, Vol. 26, DOI: 10.1590/0104-1428.2100, 2016

[ASE08] Aguado, J.; Serrano, D. P.; Escola, J. M.: "Fuels from Waste Plastics by Thermal and Catalytic Processes: A Review". Industrial & Engineering Chemistry Research, Vol. 47, pp. 7982–7992, 2008

[Con19] Conversio Market & Strategy GmbH: „Stoffstrombild Kunststoffe in Deutschland 2019", August 2020

[Con21] Conversio Market & Strategy GmbH: „Stoffstrombild Kunststoffe in Deutschland 2021", November 2022

[CAS99] Clark, J. H.; Aguado, J.; Serrano, D. P.: "Feedstock Recycling of Plastic Waste". RSC Clean Technology Monographs, The Royal Society of Chemistry, 1999

[FFC+01] Font, R.; Fullana, A.; Caballero, J. A.; Candela, J.; Carcia, A.: "Pyrolysis study of polyurethane". Journal of Analytical and Applied Pyrolysis, Vol. 58–59, pp. 63–77, 2001

[Göb22-ol] Göbelbecker, J.: „Eastman will milliardenschwere Recycling-Anlage bauen". News vom 18. Januar 2022 https://www.chemietechnik.de/anlagenbau/eastman-will-milliardenschwere-recycling-anlage-bauen-65-82-51.html, abgerufen am 05.08.2022

[Kam93] Kaminsky, W.: "Recycling of polymers by pyrolysis". Journal de physique IV Proceedings, EDP Sciences, Vol. 3 (C7), pp. 1543–1552, DOI: 10.1051/jp4:19937241, 1993

[KP22] Keilbach, F.-X.; Prizinitzki: „Mit Glykol wird der Extruder zum Turbo". Kunststoffe, Vol. 1, pp. 24–25, Carl Hanser Verlag, München, 2022

[KPS04] Kaminsky, W.; Predel, M.; Sadiki, A.: "Feedstock recycling of polymers by pyrolysis in a fluidised bed". Polymer Degradation and Stability, Vol. 85, issue: 3, pp. 1054–1050, 2004

[LST+20] Lechleitner, A.; Schwabl, D.; Schubert, T.; Bauer, M.; Lehner, M.: „Chemisches Recycling von gemischten Kunststoffabfällen als ergänzender Recyclingpfad zur Erhöhung der Recyclingquote". Österreichische Wasser- und Abfallwirtschaft, Vol. 72, pp. 47–60, 2020

[MBA+16] Miandad, R.; Barakat, M. A.; Aburiazaiza, A. S.; Rehan, M.; Nizami, A. S.: "Catalytic pyrolysis of plastic waste: A review". Process Safety and Environmental Protection, Vol. 102, pp. 882–838, 2016

[MG16] Martens, H.; Goldmann, G.: „Recyclingtechnik", Springer Fachmedien, Wiesbaden, 2016

[Möl12] Möller, R.: „Reines Synthesegas aus chlorhaltigen Altkunststoffen". In: Recycling und Rohstoffe, Band 5, pp. 905–917, Hrsg.: Thomè-Kozmiensky, K.; Goldmann, D., TK Verlag, Neuruppin, 2012

[NN01-ol] "DuPont Teijin Films launches chemical recycling LuxCR™ depolymerisation process". News & Events, Petcore-europe, https://www.petcore-europe.org/news-events/227-dupont-teijin-films-luxcr.html, 11. April 2019, abgerufen am 05.08.2022

[NN02-ol] LyondellBasell: "LyondellBasell Successfully Starts Up New Pilot Molecular Recycling Facility", https://www.lyondellbasell.com/en/news-events/corporate-financial-news/lyondellbasell-successfully-starts-up-new-pilot-molecular-recycling-facility/, abgerufen am 09.08.2022

[NN03-ol] Quantafuel: "Skive plant has started production from line three". https://www.quantafuel.com/media#/pressreleases/skive-plant-has-started-production-from-line-three-3193385, abgerufen am 09.08.2022

[NN04-ol] Cassandra Oil, "Our Process". http://cassandraoil.com/en/About-us/our-process/, abgerufen am 09.08.2022

[NN05-ol] Handerek Technologies: "Process". https://handerek-technologies.com/en/technology/process/, abgerufen am 10.08.2022

[NN06-ol] Enerkem: "Carbon-Recycling". https://enerkem.com/process-technology/carbon-recycling/, abgerufen am 11.08.2022

[QNN+17] Quicker, P.; Neuerburg, F.; Noel, Y.; Huras, A.: „Sachstand zu den alternativen Verfahren für die thermische Entsorgung von Abfällen". Studie im Auftrag des Umweltbundesamtes, ISSN: 1862–4359, 2017

[RDG17] Ragaert, K.; Delva, L.; Geem, K. van: "Mechanical and chemical recycling of solid plastic waste". Waste Management, Vol. 69, pp. 24–58, 2017

[SKP20] Seitz, M.; Klätte, M.; Pohl M.: „Evaluierung von thermisch-chemischen Depolymerisationstechnologien zur Verwertung von Kunststoffabfällen". In: Thiel, S.; Thomé-Kozmiensky, E.; Quicker, P.; Gosten, A. (Hrsg.), Energie aus Abfall, Band 17, S. 105–114, Thomé-Kozmiensky Verlag GmbH, Neuruppin 2020

[SP19] Sheel, A.; Pant, D.: "Chemical Depolymerization of PET Bottles via Glycolysis". In: Recycling of Polyethylene Terephthalate Bottles. Elsevier, 2019

[Tul19-ol] Tullo, A. H.: "Plastic has a problem; is chemical recycling the solution?". Chemical & Engineering News, October 6, 2019, Vol. 97, Issue 39, abgerufen am 05.08.2022

[ZBB07] Zia, K. M.; Bhatti, H. N.; Bhatti, I. A.: "Methods for polyurethane and polyurethane composites, recycling and recovery: A review". Reactive and Functional Polymers, Vol. 67, pp. 675–692, 2007

[ZPV+21] Zahedifar, P.; Pazdur, L.; Vande Velde, C. M. L.; Billen, P.: "Multistage Chemical Recycling of Polyurethanes and Dicarbamates: A Glycolysis-Hydrolysis Demonstration". Sustainability, Vol. 13 Iusse: 3583, pp.1–13, 2021

6 Energy Recovery

Even before material recycling, energy recovery remains the primary recovery route for plastic waste in Germany. In 2021, 64.4 % of plastic waste was recovered in waste incineration plants or as substitute fuels in order to make the energy contained in the plastic usable. According to the waste hierarchy of the Circular Economy Act (KrWG § 6, paragraph 1, see Section 2.2.1), energy recovery ranks fourth, before disposal. Thus, recycling is generally to be preferred over energy recovery. The so-called "calorific value clause" has also been invalid since 01.06.2017, which had stated that energy recovery is equivalent to material recovery if the calorific value of the waste is at least 11,000 kilojoules per kilogram. According to what criteria is energy recovery permissible at all? This question is addressed at the end of the chapter, after first presenting the key aspects of energy recovery in an overview and then the processes used for this purpose. In general, it can be said that energy recovery is an integral part of waste treatment and fulfills an important function as a pollutant sink [Spo18, p. 102].

6.1 Overview and Definitions

In energy recovery, the energy contained in plastics is recovered through combustion. The product of combustion is thermal energy, which can be used for steam and/or electricity generation or directly as process heat. This type of recovery is frequently used for plastic waste in the case of heavily contaminated and mixed fractions, for which material or feedstock recycling is not expedient, taking into account ecological and economic criteria in accordance with the applicable legal requirements.

The following objectives are pursued in the incineration of waste according to [MG16] [MG16, p. 508]:

- "Complete decomposition of organic substances into inorganic substances (so-called inertization),
- Complete oxidation of organic substances and carbon to CO_2 and H_2O (to a minor extent HCl, SO_2, NO_x),

- Maximum utilization of the released energy content in electricity and heat,
- Formation of a homogenized, greatly volume-reduced mineral combustion residue (ash, sinter or slag) with very low C content (<1 or <3 % residual carbon) as grate ash or similar,
- Suitable discharge for mechanical processing of the ash/slag into building material and for the recovery of metals [...],
- Discharge of hot flue gases for energy utilization and subsequent intensive purification with separation of fly ash."

Inertization also includes the destruction of pollutants, which are treated as harmful gases in flue gas cleaning [KC10, p. 293]. The classic, primary objectives of waste incineration are the aforementioned objectives 1, 2, 4 and 6. Energy utilization and the recovery of materials, on the other hand, are to be regarded as secondary from the perspective of waste treatment [BH13, p. 312]. For the determination of the substitution potential for other fuels in the course of a comparative assessment of this recovery route versus recycling, however, energy utilization represents an essential component.

With regard to the processes of energy recovery, after disposal of the used products, either direct incineration in waste incineration plants (WIP) or pre-treatment of the waste into substitute fuels takes place. These are then incinerated either in refuse-derived fuel power plants or in industrial plants (see Figure 6.1).

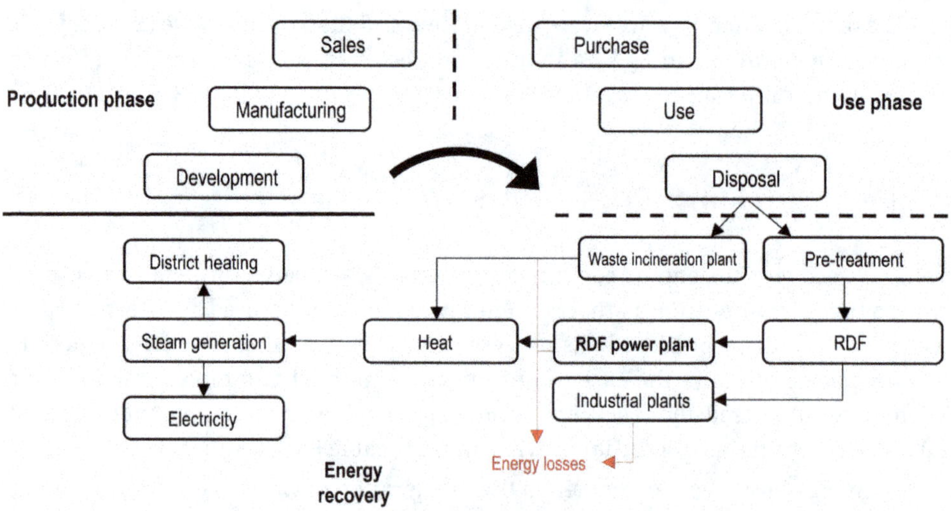

Figure 6.1 Overview of the processes in the energy recovery of plastic waste

In the case of WIPs and RDF power plants, the heat generated during combustion is used to produce steam, which can be used as district heat or for electricity production.

In contrast to material or feedstock recycling, the material cycle of plastics comes to a final end here.

From Table 6.1 it is evident that the standard plastics PE, PP and PS have similarly high calorific values as heating oil or diesel. Compared to hard coal or natural gas, they are even significantly higher. PVC or PTFE are plastics that, by contrast, exhibit significantly lower calorific values.

Table 6.1 Calorific Values of Selected Materials in Comparison, Data from [BH13][1] [MG16][2], [Car95][3], [Jun04][4], [NN01-ol][5] and [BBM+95][6]

Material	Lower calorific value HU/MJ/kg	Material	Lower calorific value HU/MJ/kg
Mixed municipal solid waste	9–11[2]	Natural gas	33–36[5]
"Teflon" (PTFE)	4.6[3]	Mixed plastic waste	ca. 35
Wood	16[1]	Polystyrene (PS)	40[3]
Paper	17[2]	Heating oil (EL/S)	42.6/39.5[5]
Polyvinyl chloride (PVC)	18[3]	Diesel fuel	42.6[5]
Polyamide (PA)	28[4]	Polypropylene (PP)	44[3]
Hard coal	30[2]	Butane	45.7[5]
Used tires	30[1]	Propane	46.3[5]
Polyethylene terephthalate (PET)	30[4]	Polyethylene (PE)	48[3]

In practice, however, no pure plastic streams are fed to a waste incineration plant, but rather mixed municipal solid waste. These have fluctuating calorific values between 9 and 11 MJ/kg. Used tyres, with a quantity of 196,000 tonnes per year, are used particularly in cement production as fuel [ZSW20, S. 47].

Of the 5.67 million tonnes of plastic waste in Germany in 2021, 64.4% or 3.66 million tonnes were recovered for energy (see Figure 6.2). With 2.13 million tonnes, the largest share goes to waste incineration plants, while the rest is used as refuse-derived fuel (RDF). The latter is processed from high-calorific waste streams and either burned in refuse-derived fuel power plants (RDF-PP) or used together with conventional fuels in power plants, in the cement or lime industry [Con21].

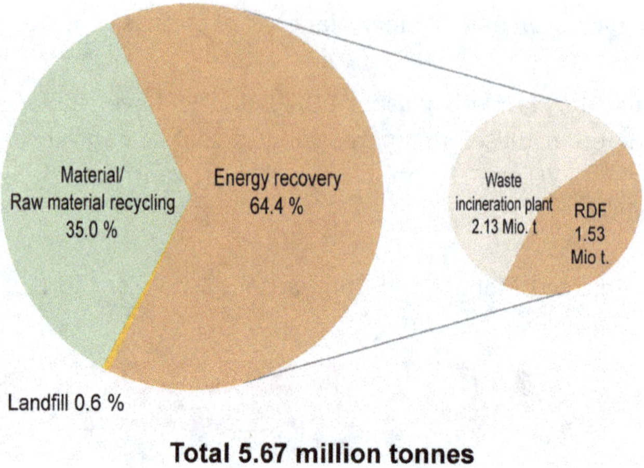

Figure 6.2 Utilization of plastic waste in Germany in the year 2021 and shares of energy recovery options, figures from [Con21]

6.2 Waste Incineration Plants

Originally, waste incineration plants were used in the context of thermal waste treatment to destroy the pollutant potential of waste and to achieve a volume reduction with regard to further disposal. Nowadays, most of these plants are energy suppliers and thus part of energy recovery, i.e. thermal treatment with energy extraction. Decisive for classification as energy recovery is the energy efficiency of the plants, which must be 0.65 (previously 0.6) for plants approved after 2008. If this value is not achieved, the so-called R1 criterion ("main use as fuel or as another means of energy generation") of the Circular Economy Act is not fulfilled. It would then be the disposal procedure D10 ("incineration on land") (Annexes 1 and 2 of the KrWG). All 66 waste incineration plants currently in operation in Germany use the energy for electricity generation and/or district heating. Six of these plants produce only electricity, all others are operated with combined heat and power. The total capacity of these plants is 20.6 million tonnes. At full capacity and after deducting the imported waste quantity of 700,000 tonnes, plastic waste accounts for only 10.8% of the recovered quantity [FHQ+18].

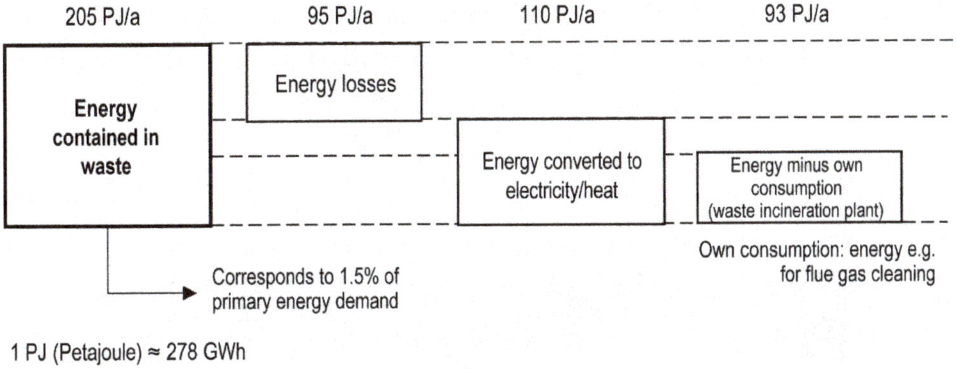

Figure 6.3 Energy conversion in waste incineration plants in Germany, figures from [FHQ+18]

The total amount of waste fed to all WIPs in Germany has an energy content of just over 205 petajoules (PJ) per year, which corresponds to approximately 1.5% of primary energy demand. With a gross utilization rate of over 50%, 110 PJ per year are converted into electricity or heat. After deducting the internal energy consumption, which is attributable among other things to the complex flue gas cleaning process, 93 PJ per year remain as the energy actually delivered by the WIPs (see Figure 6.3). Thus, the waste incineration plants contribute approximately 1% to the energy consumption required in Germany [FHQ+18].

One of the most common firing systems found in waste incineration plants is the grate firing system, the structure of which is in Figure 6.4 is shown schematically. From a waste bunker, the waste is fed to the combustion chamber using a grab arm. The step-shaped grate can be designed in different ways, but always has the task of ensuring, by means of movements, a mixing of the feed material with the primary air supplied from below and a transport of the waste to the end of the furnace.

The primary air is supplied in a temperature-controlled manner to the individual combustion zones. At the end of the furnace there is a discharge device through which the residues of the combustion process (ash, unburned components, metals, ceramics, etc.) are removed from the combustion chamber. The space above the grate represents the gasification and post-combustion zone, which will be discussed further below. This is followed by the steam boiler, which ensures energy extraction. The flue gases are partially recirculated and finally subjected to elaborate flue gas cleaning [MG16, S. 510].

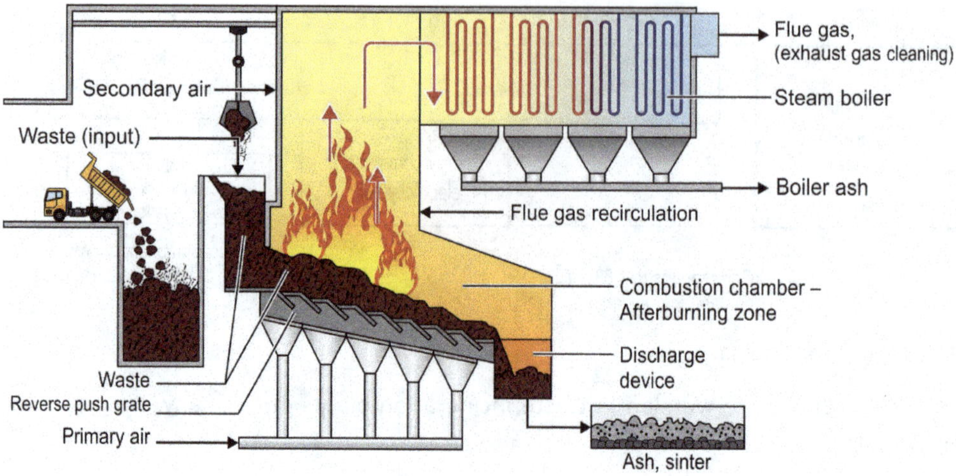

Figure 6.4 Structure of a grate firing system, according to [MG16, p. 510]

The combustion process can in principle be divided into five phases, which may also overlap within the combustion chamber (see Figure 6.5). The first phase is the drying phase at approximately 100 °C, in which the waste is freed from moisture. With additional heating in the degassing zone, the escape of volatile components follows. Still located on the grate, the burnout zone follows, in which the waste is completely combusted, down to a residual ash.

In the area of the grate, temperatures of over 1000 °C can occur. In the gasification zone above the grate, the pyrolysis gases combust, secondary air is added and exhaust gases from the plant are recirculated in order to ensure temperatures of up to 1200 °C. The post-combustion zone is connected downstream for complete combustion and for the reduction of carbon monoxide. The requirement of the Federal Immission Control Ordinance is to achieve a residence time of two seconds at a temperature above 850 °C.

After discussions about harmful emissions from waste incineration plants were still prominent in public debate in the 1990s, regulations regarding elaborate exhaust gas purification became increasingly strict [BBM+95, S. 813]. Today, plant operators must comply with the strict requirements of the 17th Federal Immission Control Ordinance, which permits lower limit values for the incineration of waste than for other combustion plants. If the latter are operated, for example, with biofuels such as wood or biogas, the emission values are many times higher than those from the incineration of waste in a waste incineration plant currently operating in Germany [MG16, S. 506 f.]. The fact that waste incineration plants are nevertheless classified by the public as environmentally harmful due to their allegedly high exhaust gas emissions therefore lacks any basis for an objectively conducted discussion.

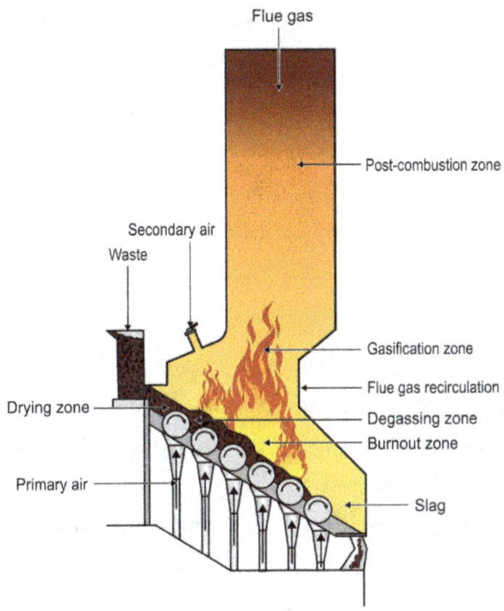

Figure 6.5 Phases of the combustion process, according to [BH13, p. 319]

6.3 Refuse-Derived Fuel Power Plants

While waste incineration plants have historically been designed primarily for thermal waste treatment, refuse-derived fuel power plants (RDF-PP) have been planned primarily for energy generation. Therefore, they were also built in location-optimized proximity to large energy consumers. Technically speaking, the plants differ little from the waste incineration plants; the substitute fuel power plants also mainly feature grate firing systems. They are designed for the use of processed substitute fuels and therefore do not have as broad a spectrum of acceptable waste materials as the waste incineration plants.

In Germany, 32 such facilities with a capacity of 6.3 million tonnes are in operation. Due to the high-calorific waste processed for the operation of the facilities, a significantly higher overall utilisation rate of approximately 70% is achieved compared to the WIPs. According to extrapolations, the waste processed in the facilities has an energy content of approximately 68 PJ per year. The energy generated amounts to approximately 44 PJ per year. It can therefore be stated that although approximately three times the amount of waste is incinerated in the WIPs, only approximately twice the amount of energy is generated compared to the RDF-PPs.

6.4 Co-Incineration in Industrial Plants and Power Plants

In addition to the classic WIPs and RDF-PPs, waste is also burned in numerous industrial plants and power stations in Germany in order to save fossil fuels. These include in particular:

- Cement and lime works
- Coal-fired power plants
- Industrial power plants

Around the year 2018, approximately 1.5 million tonnes of waste were incinerated in coal-fired power plants, whereby this amount will presumably have decreased significantly by now due to the decommissioning of many coal-fired power plants. From the same year come figures indicating 31 industrial power plants which have a permit for the incineration of waste. However, these are mostly production waste from the wood and paper industry [FHQ+18].

With a total waste quantity of 3.2 million tonnes, corresponding to an energy content of approximately 60 PJ, the use of waste in German cement plants can be considered quite significant. 34 of these facilities and two additional lime plants use, in addition to meat and bone meal, sewage sludge and paper, particularly the high calorific value waste fractions RDF, plastic mixtures and waste tyres. These are required for the extremely energy-intensive firing process of cement clinker. Cement is one of the most important binding agents in the construction industry and consists essentially of varying proportions of calcium, silicon, aluminium and iron oxide. Cement is obtained from the raw materials limestone, quartz sand and clay, which are finely ground before the firing process. In a rotary kiln, the ground material is fired at temperatures of 200 up to 2050 °C in an oxidising atmosphere. In the process, the material reaches temperatures of 1450 °C. These temperatures require a high fuel demand, which in 2018 was provided to more than two thirds by waste-based fuels. Since the combustion residues remain in the cement clinker, attention must be paid to the respective composition when selecting the waste. In terms of a functioning waste treatment and the simultaneous utilisation of the energy content, cement plants are well suited for the quantities of waste introduced there. The high temperatures and the comparatively long residence time meet all requirements of the Federal Immission Control Ordinance with regard to pollutant destruction and flue gas composition. In addition, the ash residues are utilised as materials, so that in addition to energy recovery, material recycling is also carried out [BBM+95, p. 848 f.; FHQ+18; ZSW20].

The waste-based fuels used in cement plants consist, in terms of total energy input, of just under 36% plastics, of which a quarter comes from end-of-life tires. Of the approximately 570,000 tonnes of end-of-life tires generated annually, approx. 35–40% are thermally recovered in cement plants [FHQ+18; ZSW20].

6.5 Selection of Energy Recovery Vs. Recycling

The starting point for the decision on which recovery route takes priority is, according to the Circular Economy Act, (KrWG), the protection of human beings and the environment in general. For the evaluation process, in accordance with KrWG § 6, paragraph 2, the entire life cycle of the waste must be considered and the following four criteria must be taken into account when selecting the respective recovery route:

- "The expected emissions,
- The degree of conservation of natural resources,
- The energy to be used or recovered, and
- The accumulation of pollutants in products, in waste for recovery or in products derived therefrom." [KrWG, § 6, para. 2]

It must also be taken into account whether the respective recovery route is technically feasible, economically reasonable and socially acceptable for the waste in question. The frequent objective of this specific assessment is to decide whether the recycling route preferred according to the waste hierarchy must be revised on the basis of various criteria and whether the route of energy recovery, which is ecologically more sensible in a specific case, should be pursued. The Circular Economy Act forms the basis for applying the waste hierarchy and is implemented through waste generators and holders. Under § 8 paragraph 2 KrWG, the hierarchy-compliant waste treatment for certain types of waste can be specified in more detail by statutory ordinance (examples: end-of-life vehicle ordinance, commercial waste ordinance). In addition, there are special statutory regulations for certain types of waste or product groups, which also take precedence. These include, for example, the Electrical and Electronic Equipment Act and the Packaging Act. For all other types of waste, the provisions of the Circular Economy Act are applied directly [NN01-ol, p. 1]. Even though, in addition to end-of-life vehicles and the electronics sector, different regulations apply to the largest plastic waste stream (packaging), the KrWG shows the fundamental principles by which the waste hierarchy is implemented in German legislation. Against this background, the following explanations, each of which also includes commentary relating to plastic waste, should be viewed. The guidance on the application of the waste hierarchy published by the Federal Ministry for the Environment, Nature Conservation, Nuclear Safety and Consumer Protection in 2017 makes clear on which questions a weighing of the selection of recovery methods under the KrWG should be based [NN01-ol]. The consideration of the entire life cycle of waste — from the extraction of natural resources through processing and manufacturing, marketing, use, to waste management and its consequences — forms the basis for a holistic assessment of waste impacts. Although the methodology for the weighing is not prescribed, methods such as LCA (Life Cycle Assessment) or CBA (Cost-Benefit Analysis) are described as "established" in this context. Within the framework of the guidance, the stages of the life cycle are summarized in three phases, which are presented below along with the respective questions regarding the two recovery routes of recycling and energy recovery [NN01-ol]:

Phase: Waste Recovery and Waste Management

All measures and processes that must be carried out from the handover of the waste to the end of the recovery procedure are to be considered here. With regard to the recovery procedure, these include emissions, resource and energy input, the risk of accidents and occupational health and safety. The same aspects should also be taken into account for all transport operations. Since the processes of material recycling and energy recovery are in principle well known, a comparative assessment can be made here relatively easily. Since the production of recyclate places varying quality requirements depending on the intended use, the plant-related effort depends on this, making the link with phases 2 and 3 indispensable. High-quality recycling, for example with the removal of all contaminants, odorous substances and colorants, is technologically extremely complex and generally not competitive on a large scale. If, on the other hand, only surface contamination is removed by washing processes during preparation, the plant engineering effort is considerably lower, and emissions, energy input, etc. are also correspondingly lower.

Phase: Output Assessment: Effects of Waste Recovery

Phase 2 is not concerned with the environmental impact of the recovery processes, such as the emissions from the plants (see phase 1), but rather with the substances and the energy recovered from the processes.

In the case of recycling, the quality and resource efficiency of the product are mentioned as aspects to be evaluated. Taking the quality of the recyclate into account is important; however, the question arises as to how the quality of the secondary raw material to be produced from the waste can be evaluated in a specific case. Resource efficiency refers to the proportion of the waste fed into recycling that is actually returned to the material cycle. This consideration is important from the author's perspective, because material process losses are in many cases not taken into account, or only insufficiently so. For example, the legally established recycling quotas in the Packaging Act (VerpackG, see section 2.2.9) refer to the input to recycling and not to the respective output. It must also be considered whether recycling returns pollutants contained in the waste to the cycle or not. If this cannot be ruled out, energy recovery or landfilling may be the correct treatment route.

In the case of energy recovery, the effects of energy utilization with regard to the energy contained in the waste, the efficiency and the environmental impacts of energy utilization are relevant.

Phase: Input Assessment: Phases Prior to Waste Management

In this phase, the environmental impacts are considered in comparison to the respectively substituted resource (raw materials or fuels). In the case of recycling, the product manufactured from the recyclate is put to the test. In the case of plastics, the question arises: can the negative environmental impacts be reduced in a specific case by using recyclate instead of virgin material? This question arises in particular when products are deliberately designed with larger dimensions in the event of lower mechanical material properties of the recyclates, and thus more material is processed. This increases the quantity of resources used and the energy demand. Increased rejects due to batch variations when processing recyclate must also be included in an honest analysis. Both relate to the quality of the recycling step, which naturally depends on the input stream but also on the economically feasible options.

The aim of energy recovery is to substitute conventional fuels. The combustion process is unique and irreversible for the specific waste, so that only the energetic potential of the waste can be utilized. According to the guidance of the Federal Ministry for the Environment, what is relevant here is which energy carrier (coal, gas, oil, renewable energies) is to be replaced and, where applicable, the availability of this resource.

The overall assessment should result in a comparative weighting of opposing arguments and ultimately yield the "best option" for the specific waste. This may deviate from the fundamental waste hierarchy if it best ensures the protection of human beings and the environment. Overall, the law on this basis permits a wide range of possible justifications with complex and detailed reasoning, making verification by supervisory authorities practically impossible. In the course of an evaluation on the abolition of the calorific value clause in 2016, feedback from a survey of the federal states revealed, among other things, that no cases are known in which life cycle assessments have been used in the selection of recovery methods. A statement from one of the federal states regarding the calorific value clause in force at the time indicated that, in general, the regulation of §§ 6–8 KrWG concerning the selection of recovery methods "permits or provides for the consideration of very different criteria, the application and delimitation of which are in practice only very limitedly accessible to regulatory review, ..." [ZWD+16, p. 42].

From a scientific perspective, the "life cycle thinking" underlying the KrWG should generally be applied to compare recovery routes in a specific case; however, clear framework conditions with regard to the methodology to be used are required. Life cycle assessment is initially the appropriate tool here to quantifiably compare different recovery routes. In addition to the environmental protection objective, however, human beings and their surroundings must also be protected. According to the KrWG guidance, this includes life, health, physical, mental and social well-being, and the area of "the general public and neighbors" as formulated in the Federal Immission Control Act. This also encompasses the employees of a company. Since the social aspects have not yet been captured in life cycle assessment, they must therefore be considered separately. Overall, for each type of waste or product group, quantitative compilations based on clear guidelines should serve as the basis for decision-making. The laws and ordinances that have now been enacted for many types of waste are to be seen at this point as complexity-reducing, sector-specific alternatives to the direct application of the KrWG. It must be assumed that they have a hierarchy-compliant steering effect and thus fulfill the objectives of the relevant §§ 6–8 KrWG through quota regulations, obligations for separate collection and treatment, or similar rules [ZWD+16, p. 46]. In which cases the application of the KrWG takes precedence over special rules from ordinances is a legal question and will not be discussed further at this point.

In summary, it can be said that energy recovery will continue to be a component of a realistically implementable circular economy for the foreseeable future. An important task is the pollutant sink associated with thermal treatment. Furthermore, only small incremental increases in material recovery are to be expected in the future, unless significantly more effective regulatory instruments are introduced by legislators. Energy utilization is also, against the background of increasingly scarce resources, an important, if small, building block for conserving fossil energy carriers.

References

[BBM+95] Brandrup, J.; Bittner, M.; Michaeli, W.; Menges, G.: „Die Wiederverwertung von Kunststoffen". Carl Hanser Verlag, München, Berlin, 1995

[BH13] Bilitewski, B.; Härdtle, G.: „Abfallwirtschaft". Springer-Verlag, Berlin, Heidelberg, 2013

[Car95] Carlowitz, B: „Kunststoff-Tabellen". 4. Auflage, Carl Hanser Verlag, München, 1995

[Con21] Conversio Market & Strategy GmbH: „Stoffstrombild Kunststoffe in Deutschland 2021", Oktober 2022

[FHQ+18] Flamme, S.; Hanewinkel, J.; Quicker, P.; Weber, K.: „Energieerzeugung aus Abfällen – Stand und Potenziale in Deutschland bis 2030". Umweltbundesamt, Texte 51/2018, ISSN: 1862–4359, 2018

[MG16] Martens, H.; Goldmann, G.: „Recyclingtechnik", Springer Fachmedien, Wiesbaden, 2016

[NN01-ol] Bundesministerium für Umwelt, Naturschutz, nukleare Sicherheit und Verbraucherschutz: „Leitfaden zur Anwendung der Abfallhierarchie nach § 6 Kreislaufwirtschaftsgesetz (KrWG)", Stand: 25.09.2017, https://www.bmuv.de/fileadmin/Daten_BMU/Download_PDF/Abfall-wirtschaft/krwg_leitfaden_abfallhierarchie_bf.pdf, abgerufen am: 15.08.2022

[Jun04] Jungbauer, A.: „Recycling von Kunststoffen". Vogel Verlag und Druck KG, Würzburg, 1994

[KC10] Kranert, M.; Cord-Landwehr, K.: „Einführung in die Abfallwirtschaft". Wiesbaden, Vieweg+Teubner-Verlag, 2010

[NN01-ol] Deutsche Emissionshandelsstelle: „Standardfaktoren nach Art. 31 Abs. 1 c) Monitoring-Verordnung für Emissionsfaktoren, Heizwerte und Kohlenstoffgehalte". Anhang 4 des Leitfadens zur Erstellung von Überwachungsplänen und Emissionsberichten für stationäre Anlagen, Umweltbundesamt, 2022, https://www.dehst.de/SharedDocs/downloads/DE/stationaere_anlagen/2021-2030/Ueberwachungsplan-Emissionsbericht_Leitfaden.pdf?__blob=publicationFile&v=12, abgerufen am 19.08.2022

[Spo18] Spohn, C.: „Aktuelle Situation der Thermischen Abfallbehandlung in Deutschland vor dem Hintergrund der nationalen und europäischen Entwicklungen". In: Energie aus Abfall, Band 15, Thomé-Kozmiensky Verlag, 2018

[ZSW20] Zeschmar-Lahl, B.; Schönberger, H.; Waltisberg, J.: „Abfallmitverbrennung in Zementwerken". Umweltbundesamt, Texte 202/2020, ISSN: 1862–4804, 2020

[ZWD+16] Zotz, F.; Weißenbacher, J.; Dollhofer, M.; Greßmann, A.: „Evaluation der ökologischen und ökonomischen Auswirkungen des Wegfalls der Heizwertregelung des § 8 Abs. 3 Satz 1 KrWG". Umweltbundesamt, Texte 21/2016, ISSN: 1862–4804, 2016

7 Closing Remarks

Starting from relevant reports on global environmental problems caused by the improper disposal of plastics, this book first examined environmental pollution using selected studies. Even if the study results, in their global scaling, are based on rather vague assumptions, the extent of plastic waste present in the world's oceans and the associated consequences for wildlife is, for example, enormous. It becomes clear that in many countries of the world, the introduction of a functioning waste management system would result in a significant reduction in the input of waste into the environment.

In principle, plastics, due to their property profile, are the material of choice for numerous applications and are (at least thermoplastics) recyclable in principle. However, the generally less expensive virgin material is usually available as a pure material in large quantities, with high delivery reliability and sufficient quality, so that many manufacturers prefer it over recyclate from a purely economic perspective and due to largely stable supply chains.

For the recovery of plastics from waste, fundamental prerequisites are necessary with regard to collection and sorting logistics, processing technology, the material properties to be achieved, and the availability of corresponding applications for the recyclates. Elaborate processing procedures for high-purity recyclates can in many cases only be justified by legislative requirements mandating the use of recyclate.

In Europe, massive legislative influence has set a course that paves the way for a circular economy. Recycling quotas, minimum recyclate content requirements, or design specifications lead to increased use of recyclate and the establishment of a recycling industry that operates economically for selected material streams. Those who, as actors in the value chain, are aware of their dependencies and their position relative to other actors, and who consequently understand how to assert their interests, can compete successfully. This is addressed in particular in section 2.3. By consolidating multiple roles within a corporate group, dependencies can be reduced and competitiveness increased.

Legislative requirements will in the future not only have a greater influence on material streams that have already been affected for many years, such as packaging, but also on products from the automotive sector, mechanical engineering, or the electrical industry. A competitive disadvantage compared to companies from other regions of the world that are not subject to these constraints in their markets is to be feared.

As far as the technological framework conditions are concerned, plastic waste can today be converted into single-grade material streams with a high sorting and separation quality using modern plant technology. Current research activities in the field of separating plastic composites are leading to the development of additional recyclate grades that can also be used in technical applications (see section 4.6).

Compared to material recycling, feedstock recycling still plays a subordinate role. Many processes that work well at laboratory scale fail at large-scale scaling for technical and economic reasons. The future will show whether the investments made by large corporations in recent years in the field of chemical recycling will lead to a significant output of raw materials.

Due to their high calorific value, plastics are a popular recovery route for energy generation via refuse-derived fuel power plants or waste incineration plants. Elaborate exhaust gas purification systems result in comparatively low emissions of harmful gases. Due to the waste hierarchy enshrined in the Circular Economy Act, recovery via recycling processes is in principle preferable to energy recovery. However, life cycle assessments, for example, permit decisions that deviate from this principle, the verifiability of which is currently not possible due to a lack of fundamental and overarching standards. For this reason, recovery and recycling quotas applicable to certain types of waste serve as a complexity-reducing alternative to the direct application of the Circular Economy Act.

The interplay of long-term raw material availability, effective environmental and climate protection, economic room for maneuver in a global market, and secure waste management will remain a challenging task for politics, industry, and society against the backdrop of increasing complexity in each of these areas.

Glossary

ABS	Acrylonitrile-Butadiene-Styrene
ASA	Acrylonitrile-Styrene-Acrylate
CBA	Cost-Benefit Analysis
EBA	Ethylene-Butyl Acrylate
EBS-KW	Substitute Fuel Power Plants
EPDM	Ethylene-Propylene-Diene Rubber
EPS	Expanded Polystyrene
ERDE	Erntekunststoffe Recycling Deutschland – Harvest Plastics Recycling Germany
EVA	Ethylene-Vinyl Acetate
EVOH	Ethylene-Vinyl Alcohol
EWKFondsG	Single-Use Plastics Fund Act
EWKVerbotsV	Single-Use Plastics Prohibition Ordinance
GewAbfV	Commercial Waste Ordinance
HDPE	High-Density Polyethylene
KrWG	Circular Economy Act
LCA	Life Cycle Assessment
LDPE	Low-Density Polyethylene
LLDPE	Linear Low-Density Polyethylene
PA	Polyamide
PC/ABS	Polymer blend of polycarbonate and acrylonitrile-butadiene-styrene
PC	Polycarbonate
PE	Polyethylene
PET	Polyethylene terephthalate
PMMA	Polymethyl methacrylate
PP	Polypropylene
PPWR	Packaging and Packaging Waste Regulative
PS	Polystyrene

PU/PUR	Polyurethane
PVC	Polyvinyl chloride
RDF	Refuse-derived fuel
RDF-PP	Refuse-derived fuel power plant
SAN	Styrene-acrylonitrile copolymer
TPE	Thermoplastic elastomer
TPS	Thermoplastic styrene elastomer
TPU	Thermoplastic urethane elastomer
VerpackG	German Packaging Act
VerpackDG	Packaging Law Implementation Act
WIP	Waste Incineration Plant

Index

www.ingramcontent.com/pod-product-compliance
Lightning Source LLC
LaVergne TN
LVHW080622120926
847902LV00006B/61
* 9 7 9 8 9 5 0 1 8 4 1 2 3 *